Sheffield Archaeological Monographs

13

Series Editor

John R. Collis

Sheffield Academic Press

The Future of
Surface Artefact Survey
in Europe

edited by

John Bintliff, Martin Kuna and Natalie Venclová

Sheffield Academic Press

Copyright © 2000 Sheffield Academic Press

Published by
Sheffield Academic Press Ltd
Mansion House
19 Kingfield Road
Sheffield S11 9AS
England
Tel: 44 (0)114 255 4433
Fax: 44 (0)114 255 4626

Copies of this volume and a catalogue of other archaeological
publications can be obtained from the above address or from our home page.

World Wide Web - http://www.shef-ac-press.co.uk

Typeset by Aarontype Limited
and
Printed on acid-free paper in Great Britain
by Bookcraft
Midsomer Norton, Somerset

A catalogue record for this book is available from the British Library

ISBN 1-84127-134-9

Contents

List of Figures

List of Tables

List of Contributors

Paul Barford

(Formerly) State Service for the Protection of Historical Monuments
Warsaw
Poland

John Bintliff

Faculteit der Archeologie
Rijks Universiteit Leiden
PO Box 9515
2300 RA Leiden
The Netherlands

Wojciech Brzeziński

State Archaeological Museum
Dluga 52
Warsaw
Poland

Mark Gillings

School of Archaeological Studies
University of Leicester
University Road
Leicester LE1 7RH
UK

Zbygniew Kobyliński

(Formerly) State Service for the Protection of Historical Monuments
(Currently) Institute of Archaeology and Ethnology
Polish Academy of Sciences
Solidarnosci 105
Warsaw
Poland

Martin Kuna

Archeologický ústav AVČR (Institute of Archaeology in Prague)
Letenská 4
11801 Praha 1
Czech Republic

Evžen Neustupný

Archeologický ústav AVČR (Institute of Archaeology in Prague)
Letenská 4
11801 Praha 1
Czech Republic

Claude Raynaud

Sociétés d l'Antiquité en France Méditerranéenne
UMR 154
Montpellier-Lattes
CNRS
12 Rue des Ecoles
30520 JUNAS
France

John Schofield

Monuments Protection Programme
English Heritage
23 Savile Row
London W1X 1AB
UK

Nicolas Terrenato

Department of Classics
University of North Carolina at Chapel Hill
NC 27599-3145
USA

Natalie Venclová

Archeologický ústav AVČR (Institute of Archaeology in Prague)
Letenská 4
11801 Praha 1
Czech Republic

1. Editorial Overview

John Bintliff, Martin Kuna and Natalie Venclová

Archaeological field survey has become one of the most potent tools for analyzing human prehistory and history since the advent of intensive surface survey during the last 25 years. Its ability to reveal the settlement and population history of entire landscapes has provided an unparalleled complement to traditional excavation, with its highly localized insights into past behaviour. On the other hand, the practical experience of the preceding generation of research in regional surface artefact survey has opened up a series of practical and theoretical problems that the next generation of intensive surveys must resolve. The present volume contains essays by leading experts on field survey from a wide range of European countries. They have written chapters reflecting on recent work in their country and, at the same time, were asked to look challengingly into the future of survey in their national environment.

With one exception the articles in this volume are based upon papers read at the 2nd annual meeting of the European Association of Archaeologists in Riga in 1996. The contributors participating in the session on surface artefact survey (organized by John Bintliff) were a small group and they were, of course, hardly able to bring an exhaustive description and evaluation of all the current traditions, ideas and projects within the field of surface artefact survey in Europe. We particularly regret that we were unsuccessful in eliciting contributions, either for the session or for the published volume, from some other European countries where there also exist a long tradition and much experience with field surveys. We nonetheless believe that the papers collected in this book represent a good sample of the trends and problems in this specific area of contemporary archaeological research.

The following volume illustrates surface artefact survey in both of its implicit operational aspects: as an important technique in archaeological heritage management and as a data-collecting method in academic, problem-orientated, landscape and settlement studies. To summarize general issues discussed in this volume we could perhaps focus on a number of key points that seem to emerge from the various chapters.

First, several of the papers bring valuable information on long-term field activities at the national scale. There is no doubt about the significance of such data for effective monument preservation, but the authors also show that the interpretation of the rich surface data in terms of settlement history may face various problems.

Yet we expect that there will be a general trend in the future to favour conservation and non-destructive intervention regarding archaeological sites throughout Europe; here surface artefact survey, combined with subsurface prospection deploying geophysics and geochemistry, will play a greatly enhanced role in national monument management programmes. The chapters on Poland, England and France provide important illustrations of this tendency to see surface survey as a central tool for heritage evaluation and policy formation.

Second, most of the authors share the general idea that a modern surface artefact survey must be based upon a firm theory of surface scatters and employ a sophisticated methodology in their recording and collection. Whereas even today some survey practitioners continue to express negative opinions about 'cookbook' strategies to which all future projects should adhere, it now seems increasingly necessary to lay down a series of fairly elaborate procedures for 'good practice' in surface artefact survey. This requires a commitment to a landscape-based rather than site-based field-walking methodology, to quantitative recording of a definable percentage of the surface studied, to visibility measurements, and to explicit procedures for obtaining and recording sample artefact collections from the entire landscape and from foci of discard (which should be given enhanced, gridded analytical treatment). All this will mean slower, more painstaking progress across the landscape, and larger collections so as to avoid the recurrent difficulties experienced by all recent surveys with small-number statistics (especially problematic in landscapes with many phases of occupance at contrasted levels of artefact discard and artefact survival rates). Only in this way can we make realistic comparisons between surveys, or address what are now the central interpretative issues that have emerged from the last generation of intensive surveys.

Third, during many field projects of the last two decades it has become clear that we need a better understanding of the processes bringing buried artefacts to the landscape surface, keeping them in the plough-zone, gradually changing their patterns and, quite frequently, leading ultimately to their complete disappearance. The taphonomy of the plough-zone is obviously becoming one of the critical areas for future survey research, building on pioneer experimental work and also comparisons between surface material and excavated subsurface materials at the same site (the chapter by Kuna from the Czech Republic, for example, concerns this topic). It has also become clear that careful attention must be paid to the variable conditions in which fieldwork can occur, the efficacy and adequacy of the sampling scheme and to the skills of the persons working in the field (highlighted in the chapter on Italy). Neglect of these aspects has been characteristic for most surveys until recently. This is why data from many projects bring a lot of new 'dots' to archaeological maps but very little reliable information on the settlement system of past societies. This is also why some of the authors are critical in their evaluation of the approach to field survey that still prevails in their countries.

Fourth, the intriguing results of the latest generation of intensive modern surveys show that even some very basic archaeological concepts may be questioned when the character of our evidence is changed. This has happened, for example, with the concept of the 'site', a phenomenon that is still quite understandable within an empirical approach to excavation, but which has hardly any meaning within the context of what are referred to in the present volume as 'analytical' field surveys. Considerable attention is paid to this problem in several papers, while the methodological issues, at present and for the future, relating to working not only with dot-like sites but continuous data surfaces, are described and exemplified in several chapters. This approach — the landscape as a continuous artefact — is even more desirable, if not the *only possible* one, when patterns of widely spread production activities or discard covering large segments of the countryside are being studied.

At least one contribution (from Italy) maintains that there should remain a role in the future for pioneer, extensive, survey preceding intensive survey, but it may be questioned whether the inbuilt biases and assumptions implicit in such surveys do not create harmful, artificial knowledge barriers when it comes to the second stage — the design of follow-up intensive survey programmes. Indeed, many chapters call for a (literally) 'grass-roots' approach to surface survey — the recording in the first place of the entire landsurface complement of artefacts across the zones chosen for study, only subsequent to which would various analytical methods be applied to search for structure (varied types of activity debris; cf. the chapters from Greece, England, the Czech Republic).

Fifthly, this volume illustrates the ever-increasing intervention of computer technology into the sphere of surface data analysis, synthesis and presentation. Geographical Information Systems are presented here much less as a tool but rather more as a flexible methodology that can interact dialectically with the empirical survey procedure and its results in order to bring the archaeology of surface surveys to a qualitatively higher level (see especially the chapter devoted to GIS by Gillings). But that marriage can only work on condition that the archaeological survey data have also become sophisticated enough to allow one to take advantage of such a powerful analytical tool. One of the particular developments presented in this volume (see the two chapters from the Czech Republic and that from France), and precisely challenging other surveys to follow suit to a stronger form of scientific investigation, is the combination of GIS with multivariate analysis.

The sixth and final point that emerges strongly from this collection of papers is the increasing role we envisage for Phenomenological approaches to surface artefact data (see the papers from Italy and Greece, that of Kuna from the Czech Republic, and the GIS chapter by Gillings). In contrast to some current discussion, which views landscapes as seen by past 'participant observers' as a replacement for existing survey methodologies, the contributors to this volume surely more sensibly consider the reconstruction of ancient *mentalités* as complementary and dialogic with 'outsider' analyses of past behaviours in the landscape. Terrenato also remarks importantly that the abundant artistic and literary sources for the Greco-Roman Mediterranean will provide a controlling aid to untestable explanations in this fast-growing subfield of landscape studies. We might also link this comment to the clear feeling coming through most of the chapters that future surveys cannot limit their chronological horizons to certain periods of occupance of their chosen terrain, but must give equal care to the recovery of all sequent occupances up to and including the Early Modern period. Comparison and contrast within a single landscape unit across time are at the heart of the exciting new developments in the interpretative modelling of 'community areas' or *Siedlungskammern*, discussed in both the Czech chapters and those from France and Greece.

In conclusion, it is hoped that this volume, although assembling just a limited part of what has already been done in the sphere of surface artefact survey, demonstrates the importance of surface survey projects and sets out a helpful agenda for the future design of surface artefact projects for European and other landscapes. The spatial analysis of archaeological data may yet bring new solutions, even to problems usually thought to be resolvable only by excavation. If this volume inspires new ideas of how theoretical models of past behaviour could be verified without the destructive and time-consuming intervention of excavation, its goal is achieved.

2. Beyond Dots on the Map: Future Directions for Surface Artefact Survey in Greece

John Bintliff

Summary

In 1987 Curtis Runnels and Jerry van Andel published a popular account of the results of the S.W. Argolid Survey in southern Greece, with the title *Beyond the Acropolis* (Van Andel and Runnels 1987). The title deliberately distanced the new generation of Greek field surveys, of which this was a leading example, from preceding traditions of topographic research with their emphasis on identifying the largest settlements through intuitive search procedures. Studying the entire landsurface in great intensity, looking for sites of all sizes and types, has been a hallmark of 'new wave' surveys in Greece since the late 1970s.

In many respects, however, the striking achievement of the intensive surveys of the last 20 years has been to 'put dots on the map'. Whilst the remarkable proliferation of much denser and more accurate settlement maps has been fundamental in allowing us to comprehend the geographical extent of human occupance in the different landscapes of Greece in each era of the past, as well as providing critical new information on settlement hierarchy and the extent of land utilization, the limitations of what has been achieved are becoming very clear. A series of problems has arisen that needs to be tackled by the current generation of ongoing surveys, while a renewal of methodology and interpretative approaches must seek to open up a range of issues that surveys have largely ignored, but which survey data are appropriate to address.

Among the failings and limitations of 'new wave' surveys one can list: (1) inadequate and inaccurate representation of prehistoric, early Iron Age, Early Mediaeval and 17th–18th century AD activity in Greek landscapes; (2) a failure to deal with the problem of contemporaneity of sites within phases of commonly several hundred years' length; (3) neglect of vestigial sites, obscured sites and those that appear only episodically on the surface; (4) an inability to account for the regular presence in surface scatters, alongside plentiful material of obvious 'occupation phases', of smaller amounts of artefactual material of other periods; (5) very limited work on the social and economic inferences that could be drawn from close study of surface artefact collections across entire landscapes; (6) the absence of an 'insider', or phenomenological perspective on the landscape as once experienced by ancient societies. This contribution will present these problems with examples and discuss the agenda for revitalizing Greek survey approaches to deal with current weaknesses.

Introduction

The full fruits of the New Wave of intensive Greek surveys (Bintliff 1994; Cherry 1983), a movement that commenced during the 1970s, are now appearing with the final project volumes from the Laconia, Kea, Argolid and Methana projects (Cavanagh *et al.* 1996; Cherry *et al.* 1991; Jameson *et al.* 1994; Mee and Forbes 1997), to be followed in the near future by the first fascicules of the Boeotia Project (for preliminary results see Bintliff 1991b; 1992; 1995; 1996; 1997a; Bintliff and Snodgrass 1985; 1988a; 1988b), and the final publication of the Nemea Project (preliminary results in Cherry *et al.* 1988). Their results were anticipated in the smaller-scale intensive surveys of the Agiofarango Valley in Crete (Blackman and Branigan 1977) and the Melos Survey (Renfrew and Wagstaff 1982). These 'intensive' field-walking programmes represented a marked break with the pioneer, extensive regional survey tradition that reached its acme with the University of Minnesota Messenia Survey (McDonald and Rapp 1972). Close-order fieldwalking of large areas of contiguous landscape, the recording of off-site as well as on-site information, the separation, on-site, of surface finds of different periods to gain information on the fluctuating extent of sites, are all features typical of the 'new wave' and lacking in the older, extensive tradition.

But, and especially in the Greek Lowlands, although immensely interesting and important results will now become available from this first generation of intensive survey projects, their sophisticated procedures of fieldwork, analysis and self-criticism have ironically opened up major problems for future surveys in Greece.

The first is that it is now clear that Lowland Greece is a landscape whose surface artefact cover is unusually dense and complexly patterned, and can be seen as a virtually continuous surface of human impact (Figure 2.1: a sector of some 5 sq km of open country totally fieldwalked in SW Boeotia south of the ancient city of Thespiae [open grid], reconstructed total density plot of surface artefacts in sherds per hectare, adjusted for surface visibility variations). Moreover, the dominant influence on surface arte-facts is the on-site and off-site behaviour of a handful of historic periods (especially Classical, Roman and Modern), and this tends to distort our access to the patterns left by other periods (Figure 2.2: graph of a representative sample of surface pottery from one major sector of the Boeotian landscape shown in Figure 2.1 to illustrate a localized swamping of the surface scatter through finds of the Classical era — the columns marked g-h and h).

Even if we restrict ourselves to these very periods, whose abundant ceramics litter the landscape, we find that some major advances are now required to improve our resolution. Thus the Kea Survey (Cherry *et al.* 1991), as most others, found that multiperiod sites were common. When surface collections taken away for study tend to be similar in size, this char-acteristic reduces the number of pieces for each period recovered in comparison to a single-phase site. Moreover, the pieces clearly diagnostic for each period are a mere fraction of the finds collected, given the dominance of cookware and tableware in worn condition (Figure 2.3: a typical example of the sampling problem — the Diaseli Otzia site from the Kea Survey with a wide range of finds in small numbers, only some phases, however, seen as signifi-cant for occupation — but the numerical differences are minor in the sample collected).

A more unexpected result has come from a large-scale urban survey that the Boeotia Project (co-directed by myself and Anthony Snodgrass of Cambridge University) undertook across the entire surface of the 26 ha ancient city of Hyettos (Figure 2.4: a map of the Hyettos city survey, dated surface finds of Classical-Early Hellenistic date; see Bintliff 1992 for a preliminary report). Some 670 distinct grid units (standard size 20 × 20 m) were used to record surface density and collect diagnostic samples for dating, with over 50,000 sherds being collected (of which some 13,000 were closely datable). These sherds were dominated by the phase when the town reached its greatest extent, the Classical Greek era (Figure 2.5: histogram of diagnostic pots, Classical era represented by columns G-A, G-C, G-H, A-C, A, C, C-H, A-H, H), encouraging me to set up a student project (Fuller 1996) to look for functional variation in ceramic finds across the city. The existence of at least 16 major phases of occupation across the city, from Neolithic to Early Modern, reminds us that even when, as here, 100–200 sherds were commonly collected per sample square, low-density discard periods might not be represented in the collection for a given square, although their chances would rise within a group of adjacent squares, due to huge

variations in discard rates per phase. In fact, even when we restricted ourselves to the Classical period, when surface finds were dense and ubiquitous across the entire city area, the number of sherds diagnostic for vessel shape was not great, and the variety of vessel types in use in this period was large (Figure 2.6). Hence, no general intra-urban patterning or functional zoning was detected even for the Classical era, but with three localized exceptions: a very low disposal area in the town centre where the agora or marketplace is known to have lain (from inscriptional evidence), an imbalance towards finewares on the acropolis where generally cult activities can be pre-dicted to have dominated, and finally there was the location of a sanctuary in the lower town, recently disturbed by ploughing, whose status from figurine fragments was already clear during the urban survey collection. Learning from these experiments in sur-face collection should enable us to design more robust collection strategies, but this will take more time and call for larger collections made in several stages to allow for feedback from initial processing.

Another discovery that highly intensive Greek survey has brought to light is the sheer complexity of surface artefact scatters. A total surface mapping is fundamental to most intensive survey projects, quan-titative mapping of surface finds regardless of the site-off-site distinction. Excellent justification for this approach comes from several results available using continuous landscape recording. Thus we have good evidence now that large-scale off-site scatters of a non-residental character can swamp the surface signal of small occupation sites (Figure 2.7: surface artefact density adjacent to Hyettos city, Boeotia, together with location of rural sites CN1–17). Site CN2 in this figure lies within the main agricultural plain of the ancient city and cannot be distinguished here in its immediate density impact from the high-density sherd carpet that covers the entire plain (the dark zone filling most of the lower quarter of the map). We would argue that the carpet concerned is a manuring halo of the type that can cover 1–2 km radius around town sites (cf. Wilkinson 1989: table 1). During primary fieldwalking, site CN2 was distinguished primarily by the fresh nature of its surface sherds rather than heightened density, but final confirmation awaited the internal patterning of the site when an intensive grid collection was made across its surface (Figure 2.8: gridded density count in 10 × 10 m squares for site CN2). In contrast, sites CN5 and 5 are Roman villa sites in rolling hilland with less off-site material around, so that their impact in terms of surrounding sherd discard clearly identified the locations (Figure 2.7) even before the core of those sites had been transected in fieldwalking.

It is also becoming clear from very detailed map-ping of total landscape artefact density that small rural cemeteries and vestigial farmsite scatters in poor visibility or under unhelpful cropping conditions can likewise be missed if recognition depends on a striking rise in artefacts. Qualitative and quantitative measurements across the total landscape are essential for recognizing at least some of these less obvious kinds of site, on a landsurface with dense debris of many periods. Revisiting, a vital activity that has been practised on several recent intensive projects in Greece, has also shown that a surface artefact density measurement is very conditional on varying surface conditions, producing phenomena from one visit to another over several years comparable to Graeme Barker's phrase from an Italian survey of 'sites that come on and off like traffic-lights' (Barker 1984). Quantitative measures help to provide trends in discard behaviour for the entire regional landsurface. Large-scale surface density mapping can be used to give a cumulative figure to help us see how much overall 'noise' there is in each locality, while, through chronological breakdown of finds continuously collected in transecting, we should be able to study the variable contribution to this from each period. Yet the Greek experience rules out the creation of any numerical rule for defining 'site level', whether as a global estimate or merely calibrated for each chronological phase, contrary to the Ager Tarraconensis model developed from Spain by Martin Millett (Carreté *et al.* 1995). The realities of surface variability oscillate too greatly to permit any short cut to the recognition of surface patterning based on an algorithm held to represent 'site level'. On the other hand, Millett's second model, that the supply and consumption of artefacts can vary significantly between periods, is a valuable insight that needs to be applied to the amounts of surface artefact both on- and off-site. Taphonomic considerations, which are not identified in the Tarraconensis model, can also create fundamental contrasts in the nature of surface data both between periods and between sites of the same period (see also Kuna, this volume).

Identifying site character

All of these insights and future problems are highlighted by the status of small prehistoric sites within these new surveys. Suspicions raised on earlier Italian surveys (cf. Di Gennaro and Stoddart 1982) are fully confirmed through our work on the Boeotia material: it seems likely that the prehistoric equivalent of a small historic farmstead with hundreds of sherds collectable today on the surface will be represented by a handful — maybe as little as 2–3 prehistoric sherds in a surface collection (Figure 2.9 shows the Boeotia Survey site VM2 with raw counts of surface artefact density for a grid of mainly 10×7.5 m sample units; Figure 2.10 the diagnostic sherds from a sample

collection for the Archaic-Hellenistic period at VM2, grid units mainly 30×10 m; Figure 2.11 the equivalent spread of diagnostic Late Bronze Age sherds). Revisits targeting prehistory will probably boost the number of artefacts from such sites to a score or more, but physical survival for most types of prehistoric ware is the problem. Often, as in Italy, the suggestion of prehistoric occupation comes from the surprise discovery of a few pieces in the abundant collection from a later settlement site. The obvious inference is that the occasional prehistoric sherd, seen or gathered during standard fieldwalking, even at close intervals, fails to alert surveyors to the potential existence of a small farmsite of that era, although such low densities are likely to be all we may expect to see today on the surface. In my view, unless we are dealing with tell sites where earthworks give the clue, or Aegean island sites with high-density obsidian finds, our count of prehistoric settlements in Lowland Greece is completely unrealistic, even if we take the latest most intensive surveys.

The reasons for this situation are well discussed in Martin Kuna's review of Bohemian prehistoric survey (this volume). In most of Europe the largest part of the prehistoric ceramic assemblage was not fired to very high temperatures, and was often coarse and sandy in composition. Over millennia such material, once exposed to surface weathering (plough damage and natural forces), has a low survival potential with increasing time. The case made by the Bohemian intensive survey specialists is both strong and critically important, that *surface* pottery of prehistoric eras surviving today will almost certainly have come to the ploughsoil relatively recently, out of a subsurface 'store' (a feature such as a domestic pit or decayed structure, a grave), and equally certainly cannot derive from a remnant palaeosol preserving non-site activities. The obvious inference is that prehistoric ceramic scatters in the contemporary ploughsoil mark, for the most part, vestigial settlement or burial sites, even though absolute numbers of sherds could be as few as ones and twos per locality.

Assistance might be expected from the contrastedly high survival value of lithic implements known to accompany prehistoric activity well into the later Bronze Age in Greece. We could very reasonably expect to see a much more varied picture of prehistoric use of the landscape from lithic data than that provided by pottery. Detailed research throughout Europe specializing in the documentation of later prehistoric lithic scatters has indeed shown that distinct types of assemblage and dispersal pattern can be used to identify residential areas, tool manufacture localities ('quarries') and generalized drop-zones where tools were made and discarded expediently across the working landscape (Richards 1990; Schofield 1987; 1991; this volume). However, in Greece all intensive survey teams record the difficulty for fieldwalkers to focus on ceramics in the soil at the same time as lithic finds. Recent projects have shown that most lithic finds are made by rare individuals with a special interest and knowledge of such materials. It is likely in dense ceramic artefact scatter regions that it is impossible to focus on these diverse classes of data at the same time. My experience with a lithic specialist line-walking behind the normal team showed that a dedicated lithic survey ignoring ceramics is the only way to recover good samples of lithics across the landsurface. Even obsidian, which is very different in appearance both to pottery and local cherts, fares little better than chert in its recognition. Nonetheless, current theory would suggest that the rather poor sample recovered of later prehistoric lithics should represent both 'site' or activity-focus localities and 'off-site' temporary work in the entire landscape under exploitation.

What may be done to address these serious shortcomings in the prehistoric record from intensive surveys? For prehistoric pottery evidence we will require total collection of ceramics along transects, then a return by prehistoric specialists to all locations with certain datable finds, where they may need to 'hoover' the locality to assess the nature of the context. For lithics, one walker in each transect team should be a trained lithic specialist who will ignore other classes of material. Only with such time- and personnel-consuming improvements to methodology can we begin to rectify the likely failings of the latest surveys. The same scenario appears likely for the poor showing of Early Iron Age and Early Mediaeval ceramics in Greece, where some fabrics recognized are certainly at risk from differential destruction and have low surface visibility.

Suprisingly, equal if different problems can now be recognized for the interpretation of surface artefact scatters of the major historic periods — especially Classical, Hellenistic, Roman and Late Roman. Although some intensive surveys have relied on a 'grab' collection across supposed 'sites', others have deployed a scatter of small sample units, and still others laid a grid across most or all of the 'site' and collected artefact samples from each unit. Nonetheless, all intensive surveys have produced pottery samples of a problematic character. Usually one or two periods are well-represented in the sample collection, and are easily identified as reflecting permanent domestic occupation or burial at the location. Frequently, however, other historic periods are represented in the artefact collection, but in lesser quantities. The case of the Diaseli Otzia site from the Kea Survey, introduced earlier (Figure 2.3) provides a typical example. No recent survey has provided a reasoned discussion on what these 'lesser periods' could be there for, and rather often an arbitrary cut-off is operated to remove periods from 'site' status,

often on the basis of a difference of one or two dated pieces. Since the sample collected is generally a tiny fraction of the surface scatter, the practice conveys obvious risks. Amongst the factors that require sustained investigation in this affair of the 'lesser historic periods' are the following:

(1) Sampling error: if a period reaches 'site level' on the basis of very small collection numbers, statistical error could distort actual occupation/significant use phases.

(2) Unintentional discrimination during collection: field teams may bias their sample of artefacts, perhaps because of the differential properties of ceramics of each period. Thus glazed tile is a very common feature and always collected from Archaic to Hellenistic sites, being replaced by unglazed tile for the subsequent Roman and Late Roman periods. Large sherds with black or red glaze (commonest in the Classical-Early Hellenistic period) are probably enhanced in surface collections compared to their successors, due to their apparent potential for diagnosticity. In the Late Roman period, combed ware amphorae are a highly visible and well-known diagnostic type-fossil even in tiny sherds; collections of this period are often poor in food preparation shapes and tableware.

(3) Whereas we reasonably associate a dense scatter of artefacts of one period with an activity focus, such as a permanent settlement or burial site, the occurrence of rather less, or much less, dense material of another phase is harder to decipher, since we are unprepared to state what a temporary, seasonal, or subphase only, type of site use would be characterized by. A greater attention to the functional distinctions apparent between 'major' and 'minor' components represented on sites might assist us.

(4) As noted earlier, Martin Millett's call for attention to varying pottery supply and consumption has not had any following in Greek survey hitherto. Possibly the high visibility of a particular period is mainly a consequence of a higher rate of ceramic usage per family.

In conclusion, the publication of detailed records of the finds from each site on recent intensive surveys in Greece, while encouraging the hypothesis of the importance of certain phases of historic activity in the landscape over others, has revealed a general failure to model the possible meanings of the less commonly represented ceramic periods that frequently appear in subordinate position amongst surface scatters. Future 'source-criticism' of these findings will certainly need to provide a richer set of scenarios beyond 'site' and 'not a site', in which a wider range of behaviours is modelled in surface artefact terms. It is far from inconceivable that the central hypothesis may be overturned, that is, that 'major phases' are an artefact of the way we have collected and interpreted the material.

Having admitted to significant failings in site recognition, we can point to a likely increased role for non-destructive subsurface and surface techniques in increasing our understanding of those surface sites we do recognize. Figure 2.12 shows an example, site VM70 from the Boeotia Survey, of a small, isolated Classical rural farmsite in present-day ploughland, ideal for the application of non-destructive subsurface prospection to follow-up surface artefact survey. The site was studied artefactually using the grid of 7.5 × 10 m units shown here. In addition a microtopographic survey was carried out (by M. Gillings) to identify vestigial earthwork features, and there were also geophysical and geochemical surveys. The inner and outer circles mark, respectively, the areas of densest surface ceramics and a surrounding 'halo' of lower density (but still above the surrounding off-site average). Figure 2.13 illustrates the interpretation of a resistivity survey across the same site (note that this is oriented differently from the preceding figure), from which it has been suggested we can isolate a large, rectangular farmhouse with internal subdivisions in the centre (marked with an A), plus negative-anomaly ditches on the outskirts of the site (marked with a B) delineating a farmyard enclosure (research of M. Gillings and N. Rimmington). Figure 2.14 shows the results of soil analyses undertaken in the central area of the site: abnormally high concentrations of trace element lead in the soil overlie the supposed farmhouse structure, a 'habitation' effect typical for prolonged settlement areas, while the phosphate concentrations seem to focus on the site outskirts on and near the putative ditch features (research of N. Rimmington; cf. Bintliff 1997d).

Widening the scope of interpretation

In the major historic periods where our survey evidence is less partial, the completion of so many intensive surveys has encouraged the development of comparative survey analysis. Sue Alcock's book *Graecia Capta* (Alcock 1993) has provided an excellent model that is focused on one particular period, Imperial Roman, and the evidence survey can provide for Greece as a whole during that era. In a recent study (Bintliff 1997c) I have tried to apply a series of interpretative models for the patterns revealed across time and space for rural and urban growth in Greece during Graeco-Roman times, bringing the enhanced quality of survey data to bear upon wider issues of regional dynamics. Figure 2.15 shows regional variations in demographic growth and urbanization within Iron Age Greece, utilizing the evidence of Greek regional (extensive and intensive) surveys. Figure 2.16 illustrates a few of the models that appear to have strong explanatory value for divergences in regional growth trajectories across this long era.

At such a level — identifying trends that span several centuries or even a millennium or more — survey is a strong if not the best tool archaeologists have. It explains why the study of medium-term and long-term trends has been so popular with archaeologists. In the Annaliste scheme of historical analysis, best-known from the work of Braudel (Bintliff 1991a; Braudel 1972) these are termed the 'moyenne' and 'longue durée' (Figure 2.17 shows the Annales model of history, in which the fabric of historical processes is created through the interaction of trends and events operating in parallel but over different timescales). The challenge is twofold if we wish to emulate in more detail the stimulating, integrated Annales way of studying past societies ('total history'):

First, can we detect the short-term events that are complementary to the impact of trends and processes operating on the other two levels?

Second, can we move from an outsider understanding of the longer-term development paths to that inner world of 'mentalités' — how the past populations thought of their regional worlds?

We do have some examples of a successful recovery of short-term events through survey. Thus, for example, the sack of Haliartos city by the Roman army in 171 BC is also marked by the almost complete absence on the surface of certain pottery types typical for the 2nd century BC and later (Figures 2.18 and 2.19), but the lack of reoccupation was a vital aid. At another Boeotian urban site, Askra, historic sources argue for an abandonment in early Roman times, but reoccupation and the long life of the relevant ceramic types found at the site have made it difficult hitherto to recognize this hiatus from purely archaeological surface data. As John Cherry showed some 20 years ago (Cherry 1979; Figure 2.20), if we take this problem back to the Bronze Age when most ceramic or lithic finds have coarse chronologies, then the tracing of generational site maps is probably impossible. Through clever inference he suggested that most of the Early Bronze Age 'family-farm' sites shown here were unlikely to be contemporary.

Also testing the resolution of even the latest survey methods in Greece is the problem of Dark Ages: we can suspect that often sites are not very numerous or large in times of economic and political disorder, and may be placed in unusual locations. All these considerations imply that sites in these phases may lie well off the Gaussian curves of normal locations and the normal size and density of finds. This certainly is another powerful argument for avoiding those gross sampling schemes that assume strips or quadrats across landscapes can provide representative samples. The transect sampling scheme adopted by the Greek Melos Survey (Renfrew and Wagstaff 1982, Figure 2.20), as the more recent Ager Tarraconensis survey in Eastern Spain (Carreté, *et al.* 1995), while admirably suited for documenting sites whose distribution is both ubiquitous and numerous (for example, Roman villas or Greek Early Bronze Age and Classical small farmstead distributions), suffers from an inability to record sites whose distribution is highly localized and rare (in the Spanish example mediaeval settlements were found to be absent from the area surveyed). If we return to Figure 2.7 and focus on that small sector of the field survey carried out immediately to the north of the city of Hyettos in Boeotia, we can see that within an area of 500 × 400 m continuous fieldwalking identified five discrete settlements of post-Roman date (namely CN3, 4, 8, 15 and 17). The finds from each differ, but there are overlaps between them, and current analysis suggests that in total the cluster may represent a flowing sequence from the 8th–9th centuries AD (and possibly earlier) up to the 19th century AD. Hyettos city, a mere 500 m distant from the cluster, is abandoned at the end of Late Roman times, by the end of the 7th century AD, so that, in the absence of mediaeval and post-mediaeval sites from elsewhere in the Hyettos countryside so far surveyed, these small rural sites can be taken to represent a vital database for understanding the subsequent settlement history of the Hyettos district up to modern times. The CN cluster, part or all of which could well have been missed by sample survey of the region by 10–20% arbitrary blocks or transects, is a vital argument for as close to 100% survey across groupings of contiguous 'Siedlungskammern' or communes to catch the use of such areas from any conceivable microlocation. (See the related concept of 'community area' used by Czech intensive survey; Kuna, this volume.)

One result of such intensive parish-by-parish work has been to allow confrontation with ethnohistoric data through the identification of surface survey sites with named communities in archival records. Thus, for example, it is now possible to compare the extent and density of surface ceramic finds across deserted mediaeval and post-mediaeval villages, with their known populations at set times in tax and census records. Figure 2.21 illustrates the extent of dated surface ceramics for the Early Ottoman period from surface sampling at the deserted village of Panagia (site VM4 on the Boeotia survey), which, we would hypothesize (from settlement archaeology theory), could represent over 1000 people; the archival sources provide an identical figure. Furthermore, historic sources offering information on the ethnicity of villages open up the possibility for comparing ethnic variation and material culture both in terms of the surface artefactual record and evidence of vernacular architecture (if there are standing ruined buildings or geophysical evidence for house-types). Figure 2.22 shows the ethnic attribution and population size of located villages in Boeotia for the

Ottoman census of 1466. However, intensive survey projects in Greece have generally lacked published typologies for post-Roman ceramics, so that most have been unable to produce detailed maps of mediaeval and post-mediaeval sites. In general the 'new wave' surveys have therefore added very little to our knowledge of post-Roman Greece. Current work to establish just such a typochronology is well under way (Vroom 1997), but the full effects will only come with the next generation of Greek survey.

Also part of the Annales approach to past societies, but a shared zone of interest with current Post-processual theory in archaeology, is the concern with mindscapes — trying to see surveyed landscapes through the eyes of past occupants. This is a perspective that has been little developed hitherto in Greek surveys, with the exception of a growing interest in sacred landscapes (Alcock 1993: ch. 5; cf. also Bintliff 1977: vol. 1, ch. 7 for an early application of 'sacred geography'). An example of the future direction of such work in the context of Greek survey data is the construction of visual and aural catchments relating rural sites to an urban focus (see Gillings, this volume).

Nonetheless the incorporation of what we might term a Culturalist approach into the analysis of intensive survey data, if it overprivileged the conscious, cognitive sphere of past human behaviour, would be a retrogressive step. It should not result in the neglect of the differences perceived between our outsider recording of behaviours in the landscape, and the ways past people understood their world. Much of their activities may not have been cognized, or could have been seen through a 'false consciousness'. There is still to my mind an equally valid place for analytical approaches looking for cross-cultural regularities in settlement systems, and modular developments in the size of settlement types. Figure 2.23 shows the application of a territorial model to ancient towns (filled triangles) and villages (filled circles) in Classical Boeotia, constructing through Thiessen polygons the likely exploitation zones for each nucleated focus, then comparing those cells to 'ideal' support territories of 2.5 km radius. Gaps in the distribution of centres due to the limitations of extensive survey allow predictive infilling of additional villages, encouraging future fieldwork to test the reconstructions in data-poor districts. Figure 2.24 is a general cross-cultural model developed by human geographers for the territory and population size of settlement hierarchies, which, although based on recent data, conforms with archaeological settlement systems (cf. Bintliff 1997b). The strong regularities exhibited in settlement networks of later prehistoric and ancient societies may emanate as much from human ecological and socio-biological constraints as from the conscious planning programmes of ancient communities.

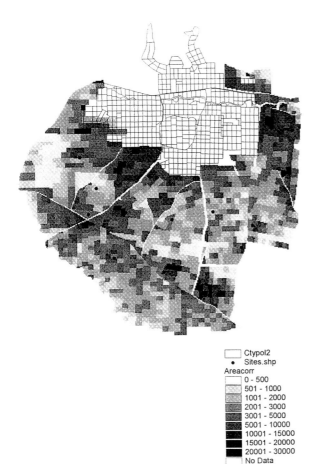

Ctypol2
Sites.shp
Areacorr
0 - 500
501 - 1000
1001 - 2000
2001 - 3000
3001 - 5000
5001 - 10000
10001 - 15000
15001 - 20000
20001 - 30000
No Data

Figure 2.1 A sector of some 5 sq km of countryside totally fieldwalked in SW Boeotia, reconstructed total density plot of surface artefacts in sherds per hectare, adjusted for surface visibility variations. City survey grid of Thespiae to north.

Diaseli Otzia Site, Kea Survey: Catalogue Entry
(Cherry *et al.* 1991: 114)

Occupation periods claimed: (Classical maybe.) Definitely Hellenistic, Late Roman and Middle Byzantine.

Dated sherds collected from the site: Archaic-Classical 2; Archaic-Hellenistic 6+; Archaic-Early Roman 1; possibly Archaic-Roman 1; Classical 2; Classical-Hellenistic 3; Classical-Late Roman 5+ (1 possible); Hellenistic 3 (and 2 possible); Late Hellenistic-Early Roman 1; Late Hellenistic-Roman 3; Late Roman 23+ (2 possibles); Roman 1; Late Roman-Modern 1; Middle Byzantine 5; Turkish-Early Modern 1; Modern 1 (1 possible). Plus 2 millstones of Classical-Hellenistic date.

Figure 2.3 A typical example of the sampling problem: the Diaseli Otzia site from the Kea Survey with a wide range of finds in small numbers within the sample collection. However, only some phases are seen as significant for occupation, but the numerical differences are minor in the sample collected.

1200
1000
800
600
400
200
0
Ceramic nos.
Series 1
preh
prehilithic
g-h
h
h-r
r-lr
lr
byz-et
t
t-mod
other
?
Ceramic phases

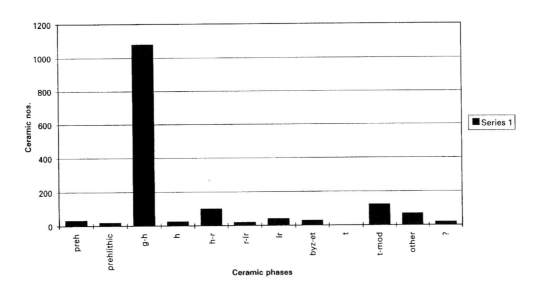

Figure 2.2 Graph of a representative sample of surface pottery from one major sector of the Boeotian landscape shown in Figure 2.1 to illustrate a localized swamping of the surface scatter through finds of the broader 'Classical' era — the columns marked g-h and h.

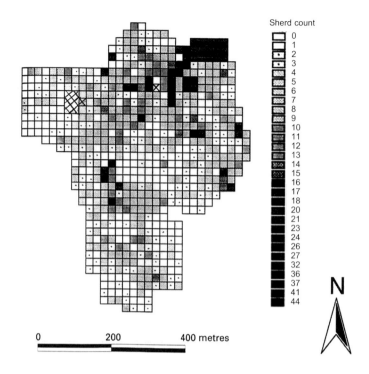

Figure 2.4   A map of the Hyettos city survey, dated surface finds of Classical-Early Hellenistic date. Standard sample grid unit 20 × 20 m.

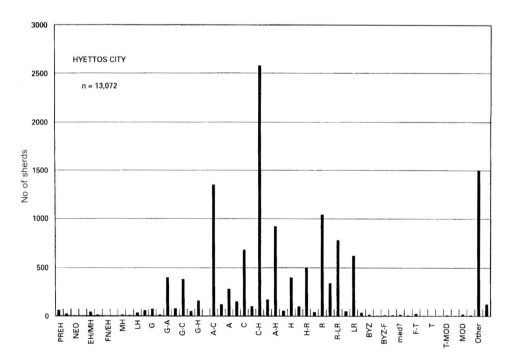

Figure 2.5   Histogram of diagnostic pottery from the sample collection derived from the city of Hyettos surface survey, the broader 'Classical' era represented by columns G–H inclusive.

Period	Code	Function	Count
c	P	basin	13
c	P	basin/mortarium	1
c	P	cooking ware	1
c	P	cooking ware?	1
c	P	cookpot	7
c	P	cookware	43
c	P	cookware, basin	1
c	P	hydria	1
c	P	hydria?	1
c	P	jar	1
c	P	jar/cooking pot	1
c	P	jug	5
c	P	krater-like basin	1
c	P	lid	1
c	P	lid?	1
c	P	mortarium	1
c	P	plainware	1
c	P/S	jar/pithos	1
c	P/S	jug/amphora	1
c	S	amphora	50
c	S	amphora?	1
c	S	cl.v.-amphora type	1
c	S	large,open shape	1
c	S	large,thick-walled vess	1
c	S	open,large vessel	1
c	S	pithos	3
c	S	pithos or tile	1
c	S	tile/pithos	2
c	SL	attic large open vessel	1
c	T	boll/skyphos	1
c	T	bowl	12
c	T	cup	16
c	T	cup/bowl	1
c	T	cup?	4
c	T	dinos	1
c	T	dish	5
c	T	giant kylix	1
c	T	kantharos	99
c	T	kantharos carinated	1
c	T	kantharos local	3
c	T	kantharos?	11
c	T	kantharos? late, small	2
c	T	kanthyros/ skyphos	1
c	T	kotyle	2
c	T	kotyle?	1

Period	Code	Function	Count
c	T	krater	8
c	T	krater, red figure	1
c	T	krater/bowl	1
c	T	krater?	2
c	T	kylix	29
c	T	kylix local	1
c	T	kylix type	1
c	T	kylix?	2
c	T	lamp	7
c	T	lamp nozzle	1
c	T	late kantharos	1
c	T	late kantharos?	1
c	T	lekane	1
c	T	lekanis	1
c	T	lekythos	1
c	T	lekythos/kantharos	1
c	T	lekythos?	1
c	T	lid or kantharos	1
c	T	miniature kantharos	1
c	T	mug	2
c	T	mug or kantharos	2
c	T	oil lamp	2
c	T	open vessel, krater?	1
c	T	palmette kylix	1
c	T	palmette skyphos	1
c	T	palmette/patterned ware	1
c	T	palmetted kylix	1
c	T	skyphos	40
c	T	skyphos/kylix	1
c	T	skyphos?	4
c	T	skypkos	2
c	T	trefoil mouth	1
c	T/P	basin/krater	1
c	T/P	bowl/basin	2
c	T/P	krater or basin	1

Period	Code	Function	Count
c	T	krater	8
c	T	krater, red figure	1
c	T	krater/bowl	1
c	T	krater?	2
c	T	kylix	29
c	T	kylix local	1
c	T	kylix type	1
c	T	kylix?	2
c	T	lamp	7
c	T	lamp nozzle	1
c	T	late kantharos	1
c	T	late kantharos?	1
c	T	lekane	1
c	T	lekanis	1
c	T	lekythos	1
c	T	lekythos/kantharos	1
c	T	lekythos?	1
c	T	lid or kantharos	1
c	T	miniature kantharos	1
c	T	mug	2
c	T	mug or kantharos	2
c	T	oil lamp	2
c	T	open vessel, krater?	1
c	T	palmette kylix	1
c	T	palmette skyphos	1
c	T	palmette/patterned ware	1
c	T	palmetted kylix	1
c	T	skyphos	40
c	T	skyphos/kylix	1
c	T	skyphos?	4
c	T	skypkos	2
c	T	trefoil mouth	1
c	T/P	basin/krater	1
c	T/P	bowl/basin	2
c	T/P	krater or basin	1

Figure 2.6   Functional variation among the surface survey sample from the city of Hyettos, the broader 'Classical' era period only (from Fuller 1996.)

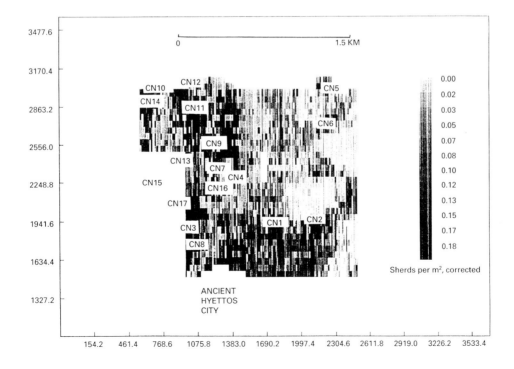

Figure 2.7   Surface artefact density adjacent to Hyettos city, Boeotia, together with location of rural sites CN1–17.

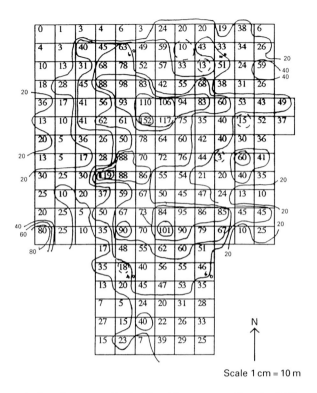

Figure 2.8   Gridded density count of surface ceramics in 10×10 m squares for site CN2.

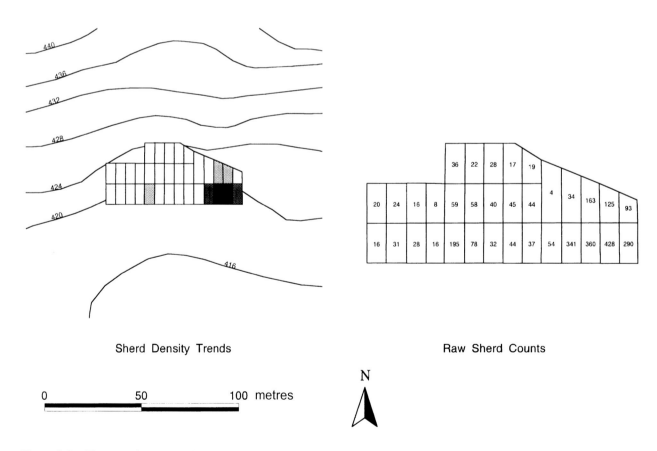

Figure 2.9   The Boeotia Survey site VM2 with raw counts of surface artefact density for a grid mainly constituted of 10×7.5 m sample units. Highest value units shaded.

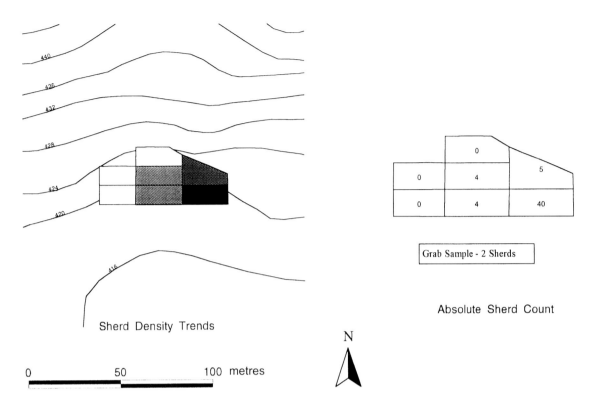

Figure 2.10 The diagnostic sherds for the Archaic-Hellenistic period from a sample collection at VM2. Grid units mainly 30 × 10 m.

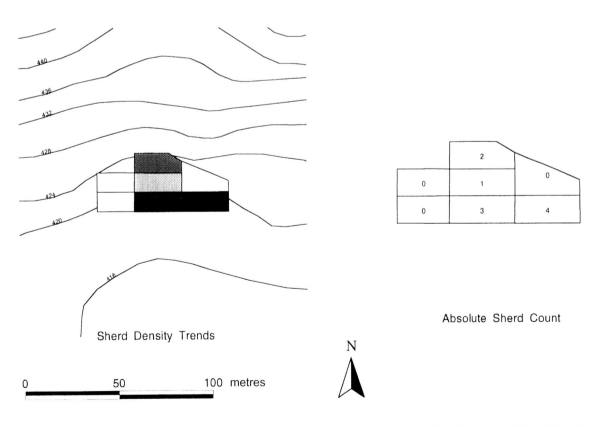

Figure 2.11 The equivalent spread of diagnostic Late Bronze Age sherds from the sample collected at VM2. Grid units mainly 30 × 10 m.

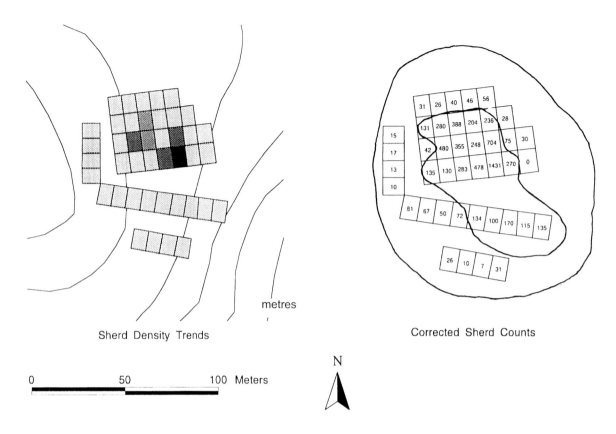

Sherd Density Trends

metres

Corrected Sherd Counts

0   50   100 Meters

N

Figure 2.12 Site VM70 from the Boeotia Survey, an example of a small, isolated Classical rural farmsite in presentday ploughland, ideal for the application of non-destructive subsurface prospection to follow-up surface artefact survey. The site was studied artefactually using the grid of 7.5 × 10 m units shown here.

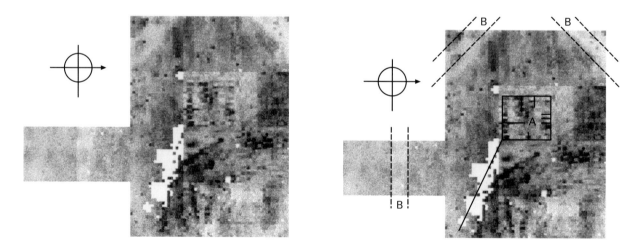

Figure 2.13 The interpretation of a resistivity survey across site VM70 (note that this is oriented differently from the preceding figure), from which it has been suggested we can isolate a large, rectangular farmhouse with internal subdivisions in the centre (marked with an A), plus negative-anomaly ditches on the outskirts of the site (marked with a B) delineating a farmyard enclosure (research of M. Gillings and N. Rimmington).

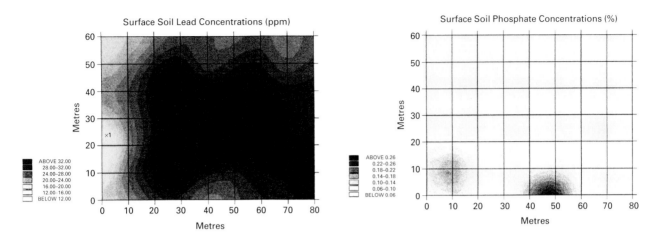

Figure 2.14 The results of soil analyses undertaken in the central area of the site VM70: abnormally high concentrations of trace element lead in the soil overlie the supposed farmhouse structure, a 'habitation' effect typical for prolonged settlement areas, whilst the phosphate concentrations seem to focus on the site outskirts on and near the putative ditch features shown in Figure 2.13 (research of N. Rimmington; cf. Bintliff 1997d).

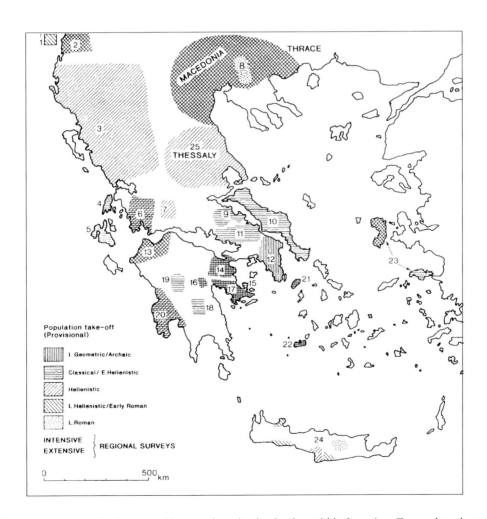

Figure 2.15 Regional variations in demographic growth and urbanization within Iron Age Greece, based on the evidence of Greek regional (extensive and intensive) surveys.

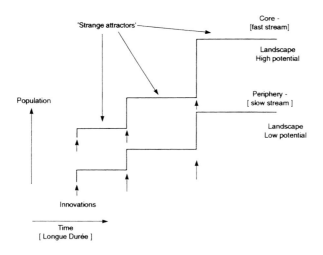

Figure 2.16 Some of the models that appear to have strong explanatory value for divergences in regional growth trajectories across Iron Age Greece.

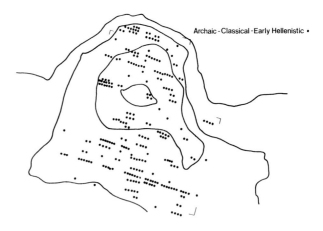

Figure 2.18 Distribution of surface ceramics of 'Classical' date from the survey of the city of Haliartos, Boeotia.

HISTORY OF EVENTS	SHORT TERM— ÉVÉNEMENTS Narrative, political history events; individuals
STRUCTURAL HISTORY	MEDIUM TERM— CONJONCTURES Social, economic history economic, agarian, demographic cycles; history of eras, regions societies; worldviews, ideologies. (*mentalités*)
	LONG TERM STRUCTURES OF THE 'LONGUE DURÉE' Geohistory: 'enabling and constraining'; history of civilizations, peoples; stable technologies, worldviews (*mentalités*)

Figure 2.17 The Annales model of history, in which the fabric of historical processes is created through the interaction of trends, events and human worldviews, operating in parallel but over different timescales.

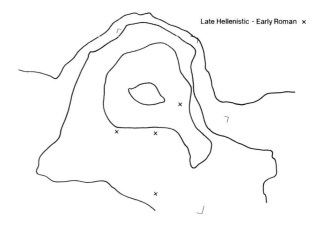

Figure 2.19: Distribution of surface ceramics of Late Hellenistic to Early Roman date from the survey of the city of Haliartos, Boeotia.

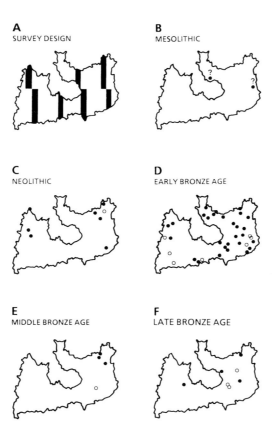

Figure 2.20 Distribution of prehistoric sites recorded by surface survey on the Aegean island of Melos, by period (Cherry 1979). • Definite ○ Probable.

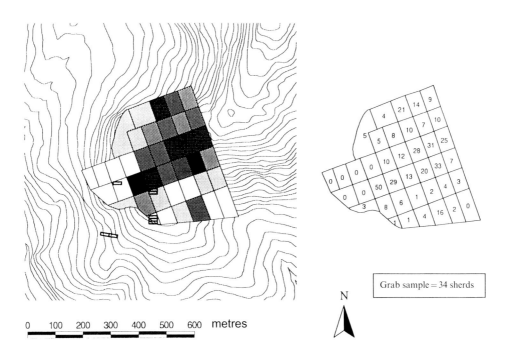

Figure 2.21 The extent of dated surface ceramics for the Early Ottoman period from surface sampling at the deserted village of Panagia (site VM4 on the Boeotia survey). Basic grid unit 50 × 75 m.

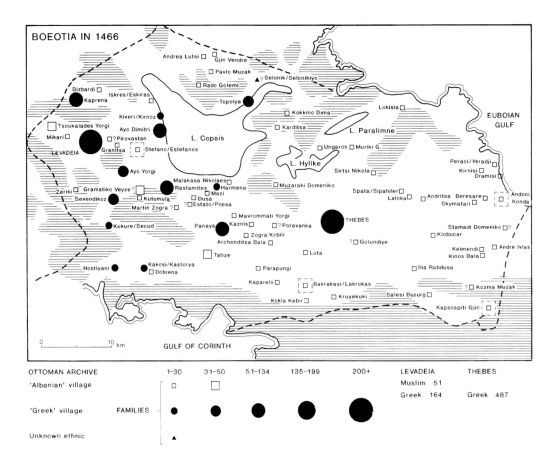

Figure 2.22 The ethnic attribution and population size of located villages in Boeotia for the Ottoman census of 1466.

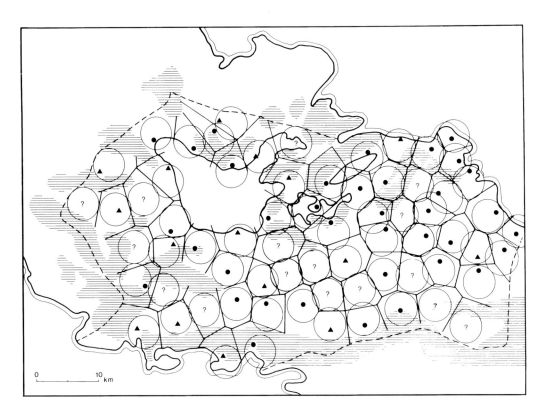

Figure 2.23 The application of a territorial model to ancient towns (filled triangles) and villages (filled circles) in Classical Boeotia, constructing through Thiessen polygons the likely exploitation zones for each nucleated focus, then comparing those cells to 'ideal' support territories of 2.5 km radius. Gaps in the distribution of centres due to the limitations of extensive survey allow predictive infilling of additional villages, encouraging future fieldwork to test the reconstructions in data-poor districts. Coastlines marked by double-lines, lakes single, upland shaded.

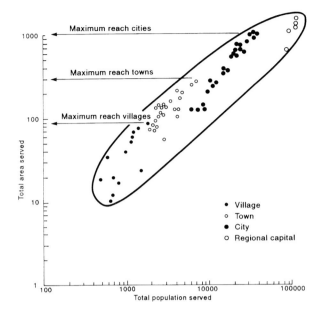

Figure 2.24 A cross-cultural model developed by human geographers for the territory and population size of settlement hierarchies, which, although based on recent data, conforms with archaeological settlement systems (cf. Bintliff 1997b). Total area served, vertical axis, in square miles.

References

Alcock, S.E.
1993 *Graecia Capta: The Landscapes of Roman Greece.* Cambridge: Cambridge University Press.
Barker, G.
1984 The Montarrenti Survey, 1982–83. *Archeologia Medievale*: 278–89.
Bintliff, J.L.
1977 *Natural Environment and Human Settlement in Prehistoric Greece.* Oxford: British Archaeological Reports, Supplementary Series 28.
1991a The contribution of an Annaliste/Structural History approach to archaeology. In J.L. Bintliff (ed.), *The Annales School and Archaeology*, 1–33. Leicester: Leicester University Press.
1991b The Roman countryside in Central Greece: observations and theories from the Boeotia Survey (1978–1987). In G. Barker and J. Lloyd (eds.), *Roman Landscapes: Archaeological Survey in the Mediterranean Region*, 122–32. London: British School at Rome.
1992 The Boeotia Project 1991: survey at the city of Hyettos. In P. Lowther (ed.), *University of Durham and University of Newcastle upon Tyne Archaeological Reports*, 23–28. Durham: University of Durham.
1994 The history of the Greek countryside: as the wave breaks, prospects for future research. In P.N. Doukellis and L.G. Mendoni (eds.), *Structures rurales et sociétés antiques*, 7–15. Paris: Les Belles Lettres.
1995 The two transitions: current research on the origins of the traditional village in central Greece.

In J.L. Bintliff and H. Hamerow (eds.), *Europe Between Late Antiquity and the Middle Ages: Recent Archaeological and Historical Research in Western and Southern Europe.* BAR International Series 617, 111–30. Oxford: Tempus Reparatum.
1996 The archaeological survey of the Valley of the Muses and its significance for Boeotian History. In A. Hurst and A. Schachter (eds.), *La Montagne des Muses*, 193–224. Geneva: Librairie Droz.
1997a The archaeological investigation of deserted medieval and post-medieval villages in Greece. In G. De Boe and F. Verhaeghe (eds.), *Rural Settlements in Medieval Europe: Papers of the 'Medieval Europe Brugge 1997' Conference 6*, 221–34. Bruges: Zellik.
1997b Catastrophe, chaos and complexity: the death, decay and rebirth of towns from Antiquity to today. *Journal of European Archaeology* 5: 67–90.
1997c Regional survey, demography, and the rise of complex societies in the ancient Aegean: core-periphery, Neo-Malthusian, and other interpretive models. *Journal of Field Archaeology* 24: 1–38.
1997d The role of science in archaeological regional surface artefact survey. In D. Dirksen and G. von Bally (eds.), *Optical Technologies in the Humanities*, 9–28. Berlin: Springer.
Bintliff, J.L., and A.M. Snodgrass
1985 The Boeotia survey, a preliminary report: the first four years. *Journal of Field Archaeology* 12: 123–61.
1988a Mediterranean survey and the city. *Antiquity* 62: 57–71.
1988b Off-site pottery distributions: a regional and inter-regional perspective. *Current Anthropology* 29: 506–13.
Blackman, D. and K. Branigan
1977 An archaeological survey of the Ayiofarango valley. *Annual of the British School at Athens* 72: 13–84.
Braudel, F.
1972 *The Mediterranean and the Mediterranean World in the Age of Philip II.* London: Fontana/ Collins.
Carreté, J.-M., S.J. Keay and M. Millett
1995 *A Roman Provincial Capital and its Hinterland: The Survey of the Territory of Tarragona, Spain.* Michigan: Journal of Roman Archaeology Supplement 15.
Cavanagh, W., J. Crouwel, R.W.V. Catling and G. Shipley
1996 *Continuity and Change in a Greek Rural Landscape: The Laconia Survey*, Vol. II. London: British School at Athens, Supplementary Volume 27.
Cherry, J.F.
1979 Four problems in Cycladic prehistory. In J. Davis and J.F. Cherry (eds.), *Papers in Cycladic Prehistory*, 22–47. Los Angeles: University of California.
1983 Frogs round the pond: perspectives on current archaeological survey projects in the Mediterranean region. In D.R. Keller and D.W. Rupp (eds.), *Archaeological Survey in the Mediterranean Area*, 375–416. Oxford: British Archaeological Reports, International Series 155.
Cherry, J.F., J.C. Davis and E. Mantzourani
1991 *Landscape Archaeology as Long-Term History.* Los Angeles: Institute of Archaeology, University of California.

Cherry, J.F., J.L. Davis, A. Demitrack, E. Mantzourani, T. F. Strasser and L. E. Talalay
  1988 Archaeological survey in an artifact-rich landscape: a Middle Neolithic example from Nemea, Greece. *American Journal of Archaeology* 92: 159–76.
Di Gennaro, F., and S. Stoddart
  1982 A review of the evidence for prehistoric activity in part of South Etruria. *Proceedings of the British School at Rome* 50: 1–21.
Fuller, S.
  1996 The Computerized Manipulation of an Urban Survey Database at Hyettos in Boeotia, Greece. Undergraduate dissertation, Department of Archaeology, Durham University.
Jameson, M.H., C.N. Runnels and T.H. van Andel (eds.)
  1994 *A Greek Countryside. The Southern Argolid from Prehistory to the Present Day.* Stanford: Stanford University Press.
McDonald, W.A., and G.R. Rapp (eds.)
  1972 *The Minnesota Messenia Expedition: Reconstructing a Bronze Age Regional Environment.* Minneapolis: University of Minnesota Press.
Mee, C., and H. Forbes (eds.)
  1997 *A Rough and Rocky Place: The Landscape and Settlement History of the Methana Peninsula, Greece.* Liverpool: Liverpool University Press.

Renfrew, C., and M. Wagstaff (eds.)
  1982 *An Island Polity: The Archaeology of Exploitation in Melos.* Cambridge: Cambridge University Press.
Richards, J.
  1990 *The Stonehenge Environs Project.* Archaeological Report 16. London: English Heritage.
Schofield, A.J.
  1987 The role of palaeoecology in understanding variations in regional survey data. *Circaea* 5 (1): 33–42.
Schofield, A.J. (ed.)
  1991 *Interpreting Artefact Scatters: Contributions to Ploughzone Archaeology.* Oxbow Monograph 4. Oxford: Oxbow.
Van Andel, T.H., and C.N. Runnels
  1987 *Beyond the Acropolis: A Rural Greek Past.* Stanford: Stanford University Press.
Vroom, J.
  1997 Pots and pans: new perspectives on medieval ceramics in Greece. In G. De Boe and F. Verhaeghe (eds.), *Material Culture in Medieval Europe: Papers of the 'Medieval Europe Brugge 1997' Conference 7*: 203–13. Bruges: Zellik.
Wilkinson, T.J.
  1989 Extensive sherd scatters and land-use intensity: some recent results. *Journal of Field Archaeology* 16: 31–46.

3. Surface Thoughts: Future Directions in Italian Field Surveys

Nicola Terrenato

Summary

The need for regional agendas specifically concerning theoretical and methodological research in field surveys has been pointed out in several recent debates. Integrated approaches, combining elements of local knowledge with worldwide advances in methods and techniques, are clearly what is called for. Italy has one of the longest traditions of field surveys in the Mediterranean region, thanks to the combined efforts of local and foreign researchers, especially British and Dutch. The recent final publication of several extensive long-term projects (such as those at Gubbio, Biferno and Venosa) sets the stage for a new assessment of the main methodological trends up to date, as well as for the definition of avenues leading to further improvement. The present paper critically reviews the recent literature, including the work done by the author in the Cecina Valley Survey. Some common trends and problems are clearly perceptible in a dynamic interaction between the British tradition and the Italian one. Future research in Italian field survey is likely to have to deal with disparate issues, ranging from the intensification of data collection procedures at the artefact level to ways of reliably reconstructing ancient landscape perception. From the point of view of methodological research, replicated coverages and simulations seem to have a considerable potential for the definition of robust research strategies and of minimum standards for future projects. On the cognitive side, the challenging relationship with textual and figurative sources will have to be explored, keeping in mind the exceptional density, richness and internal complexity that those documentary series have in peninsular Italy.

Introduction

Assessing the state of the art of field surveys in Italy is a very tall order, when one considers the richness of Italian landscapes, the amount of fieldwork carried out there over the last century and the strong geographic heterogeneity of the country. Pointing out new directions and avenues for further development is even more difficult, given that the most active, and indeed hottest debate on archaeological methodology in the country, has dealt precisely with the issue of how field surveys should be done. Considering the minimal space that theory usually has in Italian archaeology (Terrenato 1998: 175–78), it is surprising to witness how such issues as sampling in surveys fuelled an extremely active confrontation between different approaches.

When the recent literature is reviewed, a tension is clearly discernible between the intent of keeping pace with the main worldwide methodological developments and the need to adjust to local conditions and expectations. The issue of localism, as recent events are showing, is dominating Italy's public life at all levels, and, to an extent, is affecting archaeology as well. Regional agencies or individual researchers have tended to devise their own specific methodology, not always making a real effort to remain in touch with recent trends. The proliferation of regional substantive and methodological agendas thus makes the task at hand even more difficult. All the same, some traits common to the whole of peninsular Italy can be picked out, especially when we look at the past two or three decades. Thus, I will try to focus on a few main issues, namely intensification, visibility, sampling, GIS and cognitive approaches. These seem to provide the most illustrative examples of the methodological atmosphere in Italy, as well as being the areas in which future developments are to be most confidently expected.

A retrospective review

Field surveys in Italy have a comparatively long history. They originated in the early topographical work undertaken at the beginning of the century and were mainly concerned with the recording of the exceptional amount of standing structures still visible in the countryside of Italy (for a fuller historical review, see Terrenato 1996). However, systematic coverage aiming at the discovery of surface artefact scatters was only introduced by British teams during the 1950s and 1960s (Potter 1979; but esp. Hodges 1990). In parallel, or perhaps a bit later, the Italian *Forma Italiae* (a long-term series of surveys promoted by the University of Rome), while formerly mostly concerned with listing and mapping monumental remains, began recording artefact scatters in a standardized way (e.g. Quilici Gigli 1970). The Albegna and Fiora Valleys Survey (also known as the *Ager Cosanus* Survey), begun in the late 1970s and involving British,

Italian and American practitioners, perhaps best represents the methodological convergence that characterized that period (Cambi and Fentress 1988; Attolini *et al.* 1991). The following years have witnessed a sharp increase in the amount of fieldwork done, peaking in the 1980s, when a record number of projects was at work, especially in Central Italy (Mari 1983; Cucini 1985; Barker *et al.* 1986; Coltano 1986; Moreland 1987). By then, some basic methodological traits were common to both British and Italian projects: systematic coverage of large areas, site-based recording of evidence and interpretation based on the comparison of distribution maps and site counts for different phases. The methodological debate tended to concentrate on the issue of sampling, in tune with similar disputes that had previously taken place in Europe and the States (e.g. Flannery 1976: 131–36). The *Forma Italiae* and some other Italian practitioners argued strongly against gaps in coverages, while Anglo-Americans and yet other Italians, such as Carandini, viewed sampling as a convenient way of quickly obtaining a first approximation for large regions (Sommella 1989; Carandini 1989; Barker 1991). The former stressed the need for contiguous survey areas, especially if linear patterns, such as roads or centuriation, were to be grasped, while the latter were advocating the adoption of a new and more efficient technique, which, however, challenged the deep-rooted resistance to any quantitative approach that characterizes Italian archaeology (Terrenato 1998).

In the last 10 years, however, the picture has been changing considerably. The strong trend favouring a massive intensification of data collection procedures, well under way in the Mediterranean since the early 1980s (Cherry 1983), exerted a powerful influence over many survey projects in Italy. Pioneering work at Gubbio and Rieti not only took off-site evidence into account (this had already been done at Tuscania [Rasmussen 1991]) but also introduced the recording of the density of artefacts encountered in the fields (Malone and Stoddart 1994; Coccia and Mattingly 1992; 1995). Surface material was now conceptualized as a continuous distribution, with sites identified as such on the basis of precise quantitative criteria defined by the investigators. This innovative movement originated almost entirely within the circle of British scholars and brought the projects involved methodologically into line with what was happening in most of western Europe. The surprising phenomenon, however, was that Italian projects tended to remain altogether extraneous to this new trend. Survey along the coast of northern Etruria (Pasquinucci 1992), for instance, as well as new *Forma Italiae* volumes, still showed little or no interest in extra-site evidence (Mari 1991). The Cecina survey was among the few projects to be influenced by the new ideas. Its

methodology can be defined as being in an intermediate position, closer in fact to the one applied on the Tuscania survey. While all the material collected was recorded on a field-by-field basis, thus obtaining acceptable off-site information, there was no real attempt at a quantitative definition of what a site is. The identification of sites was done rather on the basis of empirical and qualitative criteria (i.e. detecting abnormal densities just on the basis of inspection or observing phenomena such as the presence of building material or other non-artefactual evidence), in contrast with straightforward density measures. Large and/or well-preserved sites were often gridded to obtain intra-site information. The deployment of energies and know-how needed for the quantification of artefact densities across the entire survey area seemed unsuitable for the goals of the project, aiming at the coverage of a large region and at the discovery of a large number of sites belonging to the 600 BC–600 AD time range (Terrenato 1992; Terrenato and Ammerman 1996).

At present, a severe methodological split seems again to be dividing the majority of Italian practitioners from foreign teams working in Italy. Methodological disagreements have always characterized field survey, and it would perhaps be a worrying sign, were they finally to be completely composed. However, what may not be a positive sign, from the point of view of a healthy dialectical archaeology (Lamberg-Karlowski 1989), is the degree of mutual exclusiveness displayed by scholarship at the moment. In general terms, by now all of the projects run by Dutch, British or American teams attempt to measure, in one way or another, the density of artefact scatters all over the sampled area, while Italian ones, with the very few exceptions mentioned above, are happy with a simpler site-based approach, with density measures applied only within the larger scatters, if at all. The most conspicuous exception to this trend is represented by the Alto-Medio Polesine-Basso Veronese project, run by a mixed British and Italian team and which has for some time been experimenting with various innovative ways of recording surface densities (Figure 3.1; Balista *et al.* 1992, with previous bibliography). This controversy is sometimes misleadingly referred to as the site/off-site debate. As the cases of Tuscania and Cecina show, off-site distribution can be recorded with only a moderate intensification of field-walking procedures; even density counts for whole fields can easily be incorporated. Where the real divide lies is in the recording of density measures for each unit of collection; in other words, in the systematic gridding (in one form or another) of the entire landscape. With this technique, sites are not simply defined in terms of presence/absence in a given field, but are located on the landscape by means of the collection units. While they are

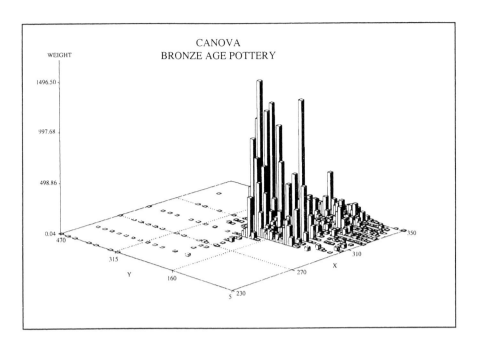

Figure 3.1   Analysis of the density of potsherds from the Alto-Medio Polesine Project (Balista *et al.* 1990: fig. 17). Each bar
            corresponds to a 5 × 5 m unit of collection in the field and expresses the aggregate weight of pottery datable
            to the Bronze Age.

ready to admit the importance of off-site scatters,
Italian practitioners are still hesitatant to take the
further methodological leap of total gridding.

As a perspective to further the debate in the future,
it can be suggested that a way forward may lie in an
honest reassessment of both methodological traditions with regard to intensity; it needs to be stressed
that the steps to be taken are not equal in size and
required effort: the reluctance to experiment with
quantitative techniques (which is not confined to
field surveys) is a trait that Italian archaeology must
by all means correct if it is to keep up with the
worldwide development of the discipline. At the same
time, methods should not simply be measured on
an absolute best practice scale, but evaluated in the
context determined by previous local knowledge,
research agendas and priorities, available resources
and expectations from the public. In this light, more
extensive approaches, especially in the case of largely
unexplored or threatened areas, may still be considered as a valuable first approximation in the investigation of a landscape (for a more detailed discussion,
see Terrenato 1996).

The disputes related above stand conspicuously in
the way of ambitions to standardize field survey
techniques, which begin to be felt in several quarters.
It has to be stated at once that a complete methodological uniformity is decidedly not a very desirable
result (Carver 1990). Indeed, a strongly dialectical
discussion is to be seen as an essential ingredient in

the progress of the discipline. However, steering well
clear from the myth of a universal best practice, it has
to be admitted that the definition of a set of minimum
requirements for carrying out field surveys would be
a major step forward, especially in the perspective of
regional syntheses based on many different survey
data sets. Some activity in this direction has been
detectable: a research network sponsored by the EU,
called POPULUS, recently provided an opportunity
to compare approaches and methodologies between
a vast range of scholars coming from different contexts. Two of the five symposia were convened by
Italian universitites, at Siena and at Pisa in 1995
(Francovich and Patterson, forthcoming; Pasquinucci
and Trément, forthcoming). The papers presented
there clearly showed the variety of methodological
approaches, and sometimes even sharp incompatibil-
ities. At the same time, the meetings had a beneficial
effect in that they brought together and prompted
interaction between scholars belonging to different
traditions. In the same year, another meeting of
Italian and British practitioners was held in Padua,
with the specific aim of discussing the feasibility of
an agreement on minimum standards for field survey procedures (Azzena, forthcoming). Dutch teams
working in Italy are coordinating their efforts and
adopting comparable methodologies (Attema *et al.*,
1998). Even the editors of the methodologically
rather conservative *Forma Italiae*, where the bulk of
the Italian field data is published, are working on a

new publication format, one that could also be adopted as a standard for some basic features of other survey reports, such as the distribution maps, the use of visibility maps, the gazetteer of sites or the publication of finds (Tartara 1999). Finally, theorists like Giovanni Azzena are making a sustained attempt to set topographical standards for the precise geographic localization of archaeological sites (Azzena and Tascio 1995).

Perspectives for the future

The current situation in Italy, as it has been reviewed above, thus seems characterized by a sharp methodological tension, as far as survey techniques are concerned. The main bone of contention, as we have seen, is represented by the issue of intensification, and any development of the discipline in the future will have to deal convincingly with it before anything else. What is meant here by intensification is the question of the level of detail in the recording of surface scatters referred to above. The present situation in Italy, as we have seen above, is that Italian and Anglo-American traditions of research are diverging sharply on this issue. To mention just one example, a lively debate on this point took place at the POPULUS meeting in Siena, with one side doubting the objectivity and reliability of site-based surveys and the other stressing the risk of getting biased results from most density-recording techniques, as well as the reputedly excessive amount of time and energy required.

The key issue here seems to be connected with assessing the impact of factors that may be involved in creating the observed distribution of artefacts: in the first place the stochastic effect of long-term ploughing, but also other potentially relevant elements, such as some recent geopedological phenomena, differential ground cover, individual variability between field walkers (and in the minimum size of sherds they collect or count), differences in the visibility of artefact classes or even in the pottery supply across time periods and in fragmentation and visibility among classes. If these are shown to be able to distort the original artefact distribution beyond recognition or interpretation (as suggested, for instance, in Fentress, forthcoming), then the actual usefulness of density measures comes into question. If, on the other hand, the results of surveys at the artefact level turn out to be consistent when repeated in different years and under different conditions, then all reluctance to accept the new technique should be abandoned. Given this situation, careful experimenting with repeated coverages seems to be a high priority and an indispensable move to obtain clear directions for the future of field surveys in Italy. It is in fact only to be expected that the impact of these potentially disturbing factors will vary locally to a great extent. In this perspective, it is essential to build up local knowledge based on experiments carried out in different parts of the peninsula and in different seasons. Once this information is acquired, simulations can also be used to help model the phenomena involved with accuracy. It would also be instructive intensively to re-survey test areas already covered by less intensive projects with the aim of assessing the differences in the reconstruction obtained. Such an approach was already part of the *Ager Cosanus* survey (which repeated an earlier coverage carried out by an American team; Dyson 1978; Cambi and Fentress 1988), and it is now included within the Tiber Valley project, which has a reassessment of the South Etruria Survey results among its goals (Patterson and Millett, 1998).

Another interesting perspective that may be relevant for the issue of intensity is now offered by the use of global positioning systmes (GPS). Experiments recently carried out at Falerii Novi by Sarah Poppy and myself (with the collaboration of Nick Ryan) seem to show that through GPS it is possible to obtain very fast and fairly accurate measures of surface artefact densities, completely eliminating the need for grids or parallel passes. On the basis of this limited experience, it would seem that the recording of surface densities could be streamlined and optimized to the point of only requiring a very modest increase in time and energy, compared to the traditional site-based Italian approach. If that turns out to be the case, then one of the main drawbacks of the artefact-level recording — that of being highly time-consuming — would disappear, breaking new ground towards a reconciliation of opposed practitioners on this very thorny issue.

The other main point where important advances are to be expected is clearly represented by visibility. Italy, as recent events help to show, is a very heterogeneous country in many respects. The one affecting survey archaeology most is the very high variability in ground cover and geopedological regime. Such phenomena as the different amount of agricultural work, the variety of crops, and the patchwork of small-scale erosion and accumulation create an extremely complex pattern of land parcels, within which archaeological sites have very different chances of being found (Verhoeven 1991). Experimental work has shown that sites may be up to ten times more common in highly visible areas when compared with poorly visible ones (Figure 3.2; Terrenato and Ammerman 1996). This means that the recovered site distribution, depending on the area, may be a very misleading reflection of the ancient landscape; consequently, a reliable recording of visibility conditions encountered at the time of the survey across the entire sampled area seems to be a very high priority for future surveys. Only with data

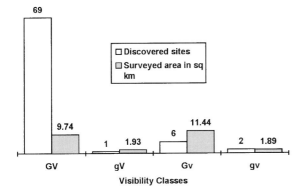

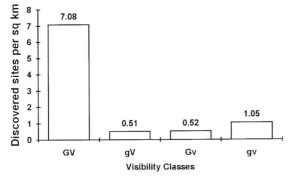

Figure 3.2   The impact of visibility at Cecina. G means favourable geopedology and g an unfavourable one, V indicates vegetation free fields and v vegetated ones. The frequency of sites found in top visibility areas (GV) is almost ten times as much as that recorded in less visible ones (Terrenato and Ammerman 1996).

of that kind can we begin to deal with the problem of the incompleteness of archaeological distributions in a realistic way. The fact that surveys recover only a fraction of the ancient distribution of sites is one that is not kept in mind enough, but which has far-reaching implications. Perhaps the most relevant is the realization that the mosaic of more, and less, visible tracts imposes an unavoidable sampling design on our coverage. This involuntary sampling cannot be simply wished away: it is there to stay and the best an archaeologist can do with it is to describe its spatial properties, so that its effects may somehow be compensated for. Simulations have had some success in experimenting with procedures that may be used to reconstruct an hypothetically complete distribution on the basis of the observed one and of the recorded pattern of visible areas (Terrenato, forthcoming).

When seen in the light of these considerations the whole controversy about sampling in field surveys (which appears to be still active in some quarters [e.g. Fish and Kowalewski 1990]) loses much of its momentum, since the recovery of complete distributions is out of the question in any case; the scope of the dilemma is thus reduced to whether to impose another man-made sampling design on top of the one

generated by nature and inherent in the landscape. This and other serious implications of visibility have not yet been fully appreciated in Italy. Another crucial one is that many commonly used techniques of spatial archaeology, such as Thiessen polygons, rank-size rule or nearest-neighbour values, are all based on the assumption that the analysed distribution is complete. Now, in most regions of Italy, it is very unsafe to take for granted completeness of recovery without a specific, local assessment of masking phenomena. To address this problem, what would really be needed is a series of case studies covering different geographic contexts of the country and evaluating the impact of visibility factors at the local level. It is to be expected that their effects will have different magnitudes according to local conditions, and the same goes for the factors potentially affecting the significance of artefact distributions mentioned above. It is becoming clear that in an heterogeneous country like Italy there is a kind of local knowledge that is as essential to the conduct of field surveys as an awareness of the worldwide developments in the discipline. The interaction of the two can lead to the definition of a set of regional methodological indications. These could result from experimental work done in accord with a regional methodological agenda, which in turn should be agreed on by practitioners interested in the same area.

Conclusions

Many other points should be touched on here for this review to even approach being exhaustive; two items, however, are perhaps particularly worthy of mention, by way of conclusion. One is represented by all the issues connected with computer data processing (and the use of GIS in archaeology in particular), and the other by the now fashionable theme of landscape perception and cognitive archaeology.

Concerning the former, it can be observed that GIS-based field surveys are becoming fairly common in Italy, even if they are used to cater for very different needs in different contexts. Artefact survey projects find it an extremely useful tool to process their quantitative data, for instance to produce density maps. In contrast, Italian projects, with their emphasis on sites and their geographic location, are mostly using GIS to build huge and topographically very precise descriptive archives. The *Forma Italiae* team, in particular, is busy trying to arrive at a storage format that can accommodate data coming from all field survey projects in Italy (Figure 3.3). Here the final aim is to obtain an electronic gazetteer of sites for the whole country, in line with similar ongoing projects in Europe (see Kobylinski *et al.*, this volume). Considerations of CRM and heritage protection are clearly prevalent in this approach, not unreasonably

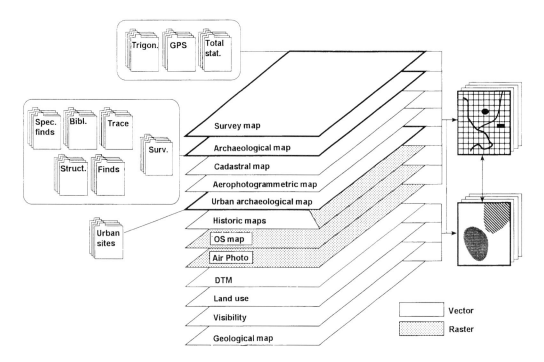

Figure 3.3   Diagram illustrating the architecture of the new GIS database that is being implemented by the *Forma Italiae* to store survey data (Azzena and Tascio 1995). The project integrates vectorial and raster cartographic layers (central stack), as well as a complex alphanumeric relational archive (left).

given the high rate of destruction by which the Italian patrimony is affected. Interestingly, these two com-plementary approaches are associated with the use of different technologies: foreign projects seem mostly to be using raster-based low-cost scientific packages like **GRASS** or **IDRISI**, while Italian ones show a preference for the large professional vector ones, like ArcInfo or Microstation, which allow the inte-gration of plans of standing structures with site distribution maps (as clearly appeared at a recent summer school organized by the University of Siena on GIS; Gottarelli 1997). In any case, however, these new tools still tend to be regarded as neater data storage systems and are only seldom used to obtain new information (see Gillings, this volume). The full heuristic value of GIS has still to be appreciated fully in Italy; its application to analyse past landscapes in innovative directions not pursuable with traditional means is clearly a high priority for future research.

The latter, 'phenomenological', issue is only begin-ning to be felt in Italy, at least in the terms in which it is treated in the foreign literature, but it may have a considerable potential, especially for the Classical and Mediaeval periods. These are actually character-ized in Italy by an exceptional density, richness and internal complexity in textual and figurative sources that may be relevant to our understanding of past landscape perceptions (Cambi and Terrenato 1994: 252 sq.). This wealth of contextual information makes the process of reconstruction much easier and more

robust: on the basis of survey data alone the risk of drifting towards pure fantasies, while attempting to recover ancient cognitive patterns, is considerable. Moreover, there are precedents in the local literature, grounded in the idealistic tradition still pervading Italian archaeology, of studies involving analyses of Roman attitudes towards the countryside or towards wild environments (e.g. Giardina 1989; Traina 1988). These are admittedly largely based on textual evi-dence, and thus almost always only relevant to elite perceptions and specific historical periods, but they can still provide a useful background for further and wider-ranging research. New studies have been build-ing on this tradition and integrating it with recent theoretical tools, such as the analysis of preferences and perceptions in various phases of occupation of the Pontine region (Attema 1994). Making good use of the highly figurative and literate character of Italy's past cultures, credible attempts can be made to explore mental maps and landscape phenomenology.

References

Attema, P.
1996   Inside and outside the landscape. *Archaeological Dialogues* 3.2: 176–95.
Attema, P., G.-J. Burgers, M. Kleibrink and D.G. Yntema
1998   Case studies in indigenous developments in early Italian centralization and urbanisation, a Dutch perspective. *Journal of European Archaeology* 1: 326–81.

Attolini, I., F. Cambi, M. Castagna, M. Celuzza, E. Fentress, Ph. Perkins and E. Regoli
1991   Political geography and productive geography between the valleys of the Albegna and the Fiora in northern Etruria. In G. Barker and J. Lloyd (eds.), *Roman Landscapes*, 142–52. (London: British School at Rome).

Azzena, G.
in press   Verso il 'minimo comune multiplo' delle ricerche di archeologia di superficie. In A. De Guio (ed.), *Archeologia di superficie: teorie e metodi a confronto*. Padua.

Azzena, G., and M. Tascio
1995   Il sistema informativo territoriale per la carta archeologica d'Italia. In M.L. Marchi and G. Sabbadini (eds.), *Venusia*, 281–97. Florence: Oschki.

Balista, C., G. Cantele, A. De Guio, M. Luciani, M. Migliavacca, R. Whitehouse and J. Wilkins
1992   Alto-Medio Polesine-Basso Veronese Project, fourth report. *Accordia Research Papers* 3: 135–62.

Balista, C., A. De Guio, M. Edwards, R. Ferri, E. Herring C. Davis, Ph. Howard, R. Peretto, A. Vanzetti, R. Whitehouse, and J. Wilkins
1990   Alto-Medio Polesine Project: second report. *Accordia Research Papers* 1: 153–87.

Barker, G.
1991   Approaches to archaeological survey. In G. Barker and J. Lloyd (eds.), *Roman Landscapes*: 1–9. London: British School at Rome.

Barker, G., S. Coccia, S. Jones, and J. Sitzia
1986   The Montarrenti Survey. Integrating archaeological, environmental and historical data. *Archeologia Medievale* 13: 291–93.

Cambi, F., and E. Fentress
1988   Il progetto topografico ager Cosanus — Valle dell'Albegna. In G. Noyé (ed.), *Structures de l'habitat et occupation du sol dans les pays Méditerranéens*, 165–79. Rome: Ecole Française de Rome; Madrid Casa de Velasquez.

Cambi, F., and N. Terrenato
1994   *Introduzione all'archeologia dei Paesaggi*. Rome: Nuova Italia Scientifica.

Carandini, A.
1989   Dibattito. In M. Pasquinucci and S. Menchelli (eds.), *La Cartografia Archeologica*, 285–90. Pisa: Provincia di Pisa.

Carver, M.
1990   Digging for data. In R. Francovich and D. Manacorda (eds.), *Lo scavo archeologico: dalla diagnosi all'edizione*: 45–120. Florence: Insegna del Giglio.

Cherry, J.C.
1983   Frogs around the pond. In D.R. Keller and D.W. Rupp (eds.), *Archaeological Survey in the Mediterranean Region*, 375–416. British Archaeological Reports, International Series 155.

Coccia, S., and D. Mattingly
1992   Settlement history, environment and human exploitation of an intermontane basin in the central Apennines: the Rieti survey 1988–1991, Part I. *Papers of the British School at Rome* 60: 213–89.
1995   Settlement history, environment and human exploitation of an intermontane basin in the central Apennines: the Rieti survey 1988–1991, Part II. *Papers of the British School at Rome* 63: 105–50.

Coltano
1986   *Terre e paduli. Reperti documenti immagini per la storia di Coltano*. Pontedera: Bandecchi e Vivaldi.

Cucini, C.
1985   Topografia del territorio delle valli del Pecora e dell'Alma. In R. Francovich (ed.), *Scarlino I: Storia e territorio*, 147–335. Florence: Insegna del Giglio.

Dyson, S.L.
1978   Settlement patterns in the Ager Cosanus: the Wesleyan University Survey 1974–76. *Journal of Field Archaeology* 5: 251–68.

Fentress, E.
in press   What are we counting for? In Francovich and Patterson (eds.) forthcoming.

Fish, S.K., and S.A. Kowalewski, (eds.)
1990   *The Archaeology of Regions*. Washington: Smithsonian Press.

Flannery, K. (ed.) 1976, *The early Mesoamerican village*. New York: Academic Press.

Francovich, R., and H. Patterson (eds.)
in press,   *Mediterranean Landscape Archaeology 5. Extracting Meaning from Ploughsoil Assemblages*. Oxford: Oxbow Books.

Giardina, A.
1989   Uomini e spazi aperti. In *Storia di Roma* 4: 71–99.

Gottarelli, A. (ed.)
1997   *Sistemi informativi e reti geografiche in archeologia: GIS-INTERNET*. Florence: Insegna del Giglio.

Hodges, R.
1990   Glyn Daniel, the Great Divide, and the British contribution to Italian archaeology. *Accordia Research Papers* 1: 83–94.

Lamberg-Karlovski, C.C.
1989   Introduction. In C.C. Lamberg-Karlovski (ed.), *Archaeological Thought in America*, 1–16. Cambridge: Cambridge University Press.

Malone, C., and S. Stoddart
1994   *Territory, Time and State: The Archaeological Development of the Gubbio Basin*. Cambridge: Cambridge University Press.

Mari, Z.
1983   *Tibur III*. Florence: Olschki.
1991   *Tibur IV*. Florence: Olschki.

Moreland, J.
1987   The Farfa Survey: a Second Interim Report. *Archeologia Medievale* 14: 409–18.

Pasquinucci, M.
1992   Ricerche topografico-archeologiche in aree dell'Italia settentrionale e centrale. In M. Bernardi (ed.), *Archeologia del Paesaggio*, 525–43. Florence: Insegna del Giglio.

Pasquinucci, M., and Ph. Trément (eds.)
in press,   *Mediterranean Landscape Archaeology 4. Non-Destructive Techniques Applied to Landscape Archaeology*. Oxford: Oxbow Books.

Patterson, H. and M. Millett
1998   The Tiber Valley Project. *Papers of the British School at Rome* 66: 1–20.

Potter, T.W.
1979   *The Changing Landscapes of Southern Etruria*. London: Elek.

Quilici Gigli, S.
1970   *Tuscana*. Rome: De Luca.

Rasmussen, T.
1991   Tuscania and its Territory. In G. Barker and J. Lloyd (eds.), *Roman Landscapes*, 106–14. London: British School of Rome.

Sommella, P.
1989   Conclusioni. In M. Pasquinucci and S. Menchelli (eds.), *La cartografia archeologica*: 291–305. Pisa: Provincia di Pisa: Privincia di Pisa.

Tartara, P.
1999   *Torrimpietra*. Florence: Olschki.

Terrenato, N.
1992   La ricognizione della Val di Cecina. In M. Bernardi (ed.), *Archeologia del paesaggio*, 561–65. Florence: Insegna del Giglio.

1996   Field survey methods in Central Italy (Etruria and Umbria). Between local knowledge and regional traditions. *Archaeological Dialogues* 3.2: 216–30.

1998   Fra tradizione e trend. Gli ultimi venti anni (1975–1995). In M. Barbanera, *L'archeologia degli italiani*, 175–192. Rome: Editori Riuniti.

Terrenato, N.
in press   The visibility of artefact scatters and the interpretation of field survey results: towards the analysis of incomplete distributions. In Francovich and Patterson forthcoming.

Terrenato, N., and A.J. Ammerman
1996   Visibility and site recovery in the Cecina Valley Survey, Italy. *Journal of Field Archaeology* 23: 91–109.

Traina, G.
1988   *Paludi e bonifiche nel mondo antico*. Rome: Nuova Italia Scientifica.

Verhoeven, A.
1991   Visibility factors affecting artefact recovery in the Agro Pontino Survey. In A. Voorrips, S.H. Loving and H. Kamermans (eds.), *The Agro Pontino Survey Project*, 87–97. Amsterdam: University of Amsterdam.

4. Surface Artefact Studies in the Czech Republic

Martin Kuna

Summary

This paper deals with general problems of surface artefact survey. As the point of departure, the concept of the site and its applicability in landscape surveys is questioned. Instead of the site-oriented methodology, an 'analytical' approach is recommended, considering artefact density over the landscape surface. The validity of 'analytical' data is, however, highly dependent upon understanding formation processes affecting archaeological assemblages in the plough-zone and on its surface. Several factors explaining variability in the density of prehistoric pottery fragments are mentioned here. To show how 'analytical' data can be exploited, the surface distribution of pottery fragments discovered by the ALRB project is used to study the continuity of residential areas over the long-term development of agricultural prehistory and the Early Middle Ages. Multivariate analysis and geographical information systems are employed for this purpose. At the end of the paper are tentatively summarized the main directions and tasks for the future development of surface survey.

Introduction

Methods of surface artefact survey in Czech archaeology have changed substantially over the last hundred years. The popularity of artefact collection as an archaeological field technique, the means of its application and the overall evaluation of surface data were, however, always connected to more general aspects of archaeological theory and practice. Hence, several stages can be traced in the development of surface studies, corresponding to the development of archaeological paradigms and, logically, to the changing objectives laid down for surface surveys.

At the beginning, the main objective was rarely anything other than artefacts themselves. Surface finds were mainly acknowledged for their *typological attributes*, contributing to knowledge of the formation and spread of archaeological cultures. Many theoretical studies already used surface finds by the second half of the 19th century, but the prevailing approach to surface data was as a rule very selective. Usually, only the most distinctive artefacts were collected and considered, stone axes being a typical example.

The acknowledgement of the *contextual value* of surface finds developed later, but still relatively early. Surface finds started to be evaluated not only for their formal qualities, but also for their ability to indicate other, more complex archaeological entities hidden below the surface. During the first half of the 20th century, surface survey developed into one of the most important methods of *archaeological reconnaissance*. This approach has, consequently, brought some improvements in survey methods (e.g. more spatial accuracy is required if the discovered sites are to be located again), but has still provided a very narrow perspective for the employment of surface data. Such a perspective has, however, survived until recently. The most recent monograph on the prehistory of Bohemia (Pleiner and Rybová 1978:20), for example, still suggests:

> Modern archaeology aims at the registering of all the chance finds which are continually coming to light ... but it also seeks for more instances ... which could help us to solve important questions about the past. This is why survey and field walking campaigns are organised to build up site records in various regions. At places identified by field walking survey an excavation can start. Opening the trenches it becomes clear where the sub-surface structures are situated and which parts of the site are sterile (translation by M.K.).

The long-lasting inability of Czech archaeology to conceptualize surface survey beyond the limits of a site reconnaissance method clearly reflected a generally very low level of interest in the non-typological aspects of the archaeological record. Most professional archaeologists have been preoccupied with chronological questions and therefore preferred excavations, the larger the better, as the only legitimate means of data collection. Surface surveys can perhaps, according to these opinions, provide some useful preliminary information, but hardly any 'proper' data contributing to the solution of the 'serious' problems of the discipline. Surface survey has usually been recommended to amateur archaeologists (Buchvaldek 1965; Vencl 1968).

This sort of underestimation of surface data was also one of the reasons why Czech archaeology almost entirely neglected opportunities raised during the period when many new surface scatters were at hand in the landscape, resulting from changes in agricultural techniques and land use in the 1950s and early 1960s. The impact that this potential source of information could have had for archaeology can be seen from the Danish example (Thrane 1989), which in the same period showed a rapid development in settlement archaeology due to the sudden increase of data from surface surveys. In the territory of the Czech Republic, very few attempts have been made to carry out surface surveys on a regional basis. If done, they have mostly been organized by amateurs as 'one-man projects', with no financial budgets and little official support (Knor 1954; Hammer 1964; 1966; Sedláček 1967a; 1967b; Fencl 1975; Kolbinger 1995). Large-scale, ambitious projects, comparable to the Polish Archaeological Record (AZP; Barford *et al.*, this volume), have never been started or considered in Bohemia and Moravia (Klápště 1985).

Despite this, a new perspective for surface artefact surveys arose in the 1960s, when several pioneering studies appeared to challenge the framework of typological archaeology. Questions regarding settlement patterns, ecology, subsistence strategies and demographic trends were now defined as important research targets (e.g. Kudrnáč 1961; Neustupný 1965; Soudský 1966). Consequently, not only isolated sites but *regional settlement patterns* became the objective of various archaeological field projects. Since the 1970s, this trend has been enhanced by the threat to the historic landscape, emerging drastically in some industrial areas, such as the open-cast coal-mining region of north-west Bohemia. Archaeologists here have been suddenly confronted with extensive landscape destruction, exceeding anything known before. To cope with this threat, new theoretical and methodological guidelines had to be developed: principles of regional research (Smrž 1986; 1987), sampling procedures (Neustupný 1984) and also a more frequent employment of the various sorts of field survey.

Amongst other issues, the concept of the *micro-region* was articulated at the end of the 1970s. Originally, the term 'micro-region' meant a small sample territory to be intensively studied instead of a whole threatened territory, which is too large to be studied effectively. Quite soon, however, this concept was underpinned theoretically, and it received the meaning of a definable geographical and cultural unit where various aspects of past cultural systems can be studied. Owing to this, a series of microregional studies were undertaken in the following period, both in regions under immediate industrial threat and outside them. These projects were based either on the

results of rescue excavations (Smrž 1986; 1987; Beneš and Koutecký 1987), long-term research excavations combined with fieldwalking (Pleinerová and Muška 1981), or, increasingly, entirely on surface data collection (Rada 1987; Meduna and Černá 1991; Břicháček and Košnar 1987; Kuna 1991a; Frolík and Sigl 1995). The targets of these studies were, for example, the definition of the size of the communities and their home territories in the Neolithic, Bronze Age and Iron Age (Beneš and Koutecký 1987; Kuna 1991a), the dynamics of residential areas in the Bronze Age (Smrž 1986; 1987) and Roman Period (Břicháček and Košnar 1987), or the colonization of the country during the Early Middle Ages (Meduna and Černá 1991). In spite of some critical remarks (Neustupný 1982; 1993b), surface survey has been gradually re-evaluated, and has metamorphosed into a crucial technique of *primary data collection* in (prehistoric) archaeology.

At the same time, surface survey has been fully adopted by some museum archaeologists and applied within their regional data acquisition strategies. In several cases, these efforts brought an enormous influx of new data. They also deeply changed views on settlement density and development in those parts of the country that were previously considered to be archaeologically poor or sterile (such as some regions of South Bohemia; see, e.g., Fröhlich and Michálek 1989).

A similar trajectory may also be suggested for the study of mediaeval settlement in Bohemia, broadly corresponding to developments in German *Siedlungsarchäologie* (Jahnkuhn 1955; 1976) and British landscape archaeology over the same timespan. A long-term programme of mediaeval settlement research was articulated in the 1960s, and further developed later (Smetánka 1970; Smetánka and Škabrada 1975). Transformations of the settlement pattern during the Early Middle Ages and particularly during the transition from the Early to High Middle Ages (12–13th century AD) have become one of the main research topics. Surface survey has been employed here as an integral part of field methodology (e.g. Frolík and Sigl 1995), able in itself to provide relevant data. This fact also led to its theoretical reconceptualization (Klápště and Žemlička 1979: 903):

> 'The density of data at hand in most regions ... is not sufficient for studies of ... settlement patterns and their changes in time. In such a situation, surface survey, traditionally seen as just a preparatory technique, should be re-evaluated and become a legitimate research method' (translation M.K.).

During the 1980s and early 1990s the evidence from regional surveys has also made clear that some of the most common concepts of the traditional paradigm may not work well when used in this context (Kuna 1991b). Intensive surveys have brought a completely new type of data that has not been fully encountered before. Traditional archaeological maps with a few isolated dots as sites have suddenly changed into dense data surfaces, which can hardly be structured in terms of spatially discrete sites. It has become clear that past human activities were usually not limited to isolated, empirically definable loci ('sites'), but rather occupied large continuous areas and created complex palimpsests of remains.

This characteristic of the data corresponds to what was foreseen by *the settlement area (community area) theory* (Neustupný 1986). According to this theory, we should be aware of the whole range of activities carried out by past communities within their home territories (community areas), and consider them within an explicit spatial model derived from the behavioural rules of living cultural systems. When matching such a conceptual model with what can be observed in the record, various kinds of *archaeological transformations* (like accumulation, reduction, etc.) must be considered (Neustupný and Venclová 1996). Some activities are necessarily invisible in the record, although they must of necessity have occupied some real space in the past, while other activities may be recognizable just from a few and/or indistinct culture traits. This theory has created a new pool of interpretative resources for the data obtained from the surface. Its contribution may, perhaps, be partly reminiscent of the issues raised by 'off-site archaeology' but it is, in fact, substantially different in several aspects, being primarily based not on empirical facts but rather on a deductive model of living culture.

To meet these goals, it is necessary to modify the methodology of surface artefact survey in several aspects. The common approach — mapping particular find concentrations in the landscape (sites) cannot bring appropriate results. Instead of 'sites', the proper and unambiguous definition of which is impossible, the *artefact density* on the ground surface should be the variable identified and measured by survey. However complicated it may be, the artefact density on the landscape surface stands in some relation to the allocation, function/meaning and intensity of past human activities, the identification of which is, after all, the main objective of modern surface surveys. My aims here are (1) to discuss the questionable concept of 'site' in the context of field surveys, and to present an *analytical method* of survey trying to avoid this concept; (2) to discuss the most important depositional and post-depositional processes in the Central European landscape, these being crucial for the plausibility of data obtained by analytical survey; and (3) to present an example of a mathematical

method for the identification of patterns in surface data and the visualization of the results by a GIS-type software.

An analytical approach to surface survey

The concept of 'site' is deeply rooted in the typological background of the culture-history (typological) paradigm in archaeology, and its empirically oriented field methodology. It also stands at the very centre of the theoretical framework of most of the surface surveys carried out in the Czech Republic to date. Generally, sites are understood to be empirically recognizable spatial units of (arte)facts, usually occurring in high densities. Since 'sites' produce most of the (typological) information used by culture-history archaeology, they are (were) logically believed to be the natural units of the archaeological record, as well as units of past human behaviour (Vencl 1995). The application of the site concept, however, meets serious problems whenever used to structure data obtained in a limited region by surface artefact collection, or any other type of intensive landscape survey.

The first problem encountered in the concept of site is the very fact that there are no objective criteria to define what a 'site' really is. Observed from the surface, prehistoric and mediaeval artefacts are often scattered over the landscape, reaching a total extent of even hundreds of hectares. They show, of course, quantitative peaks at some locations, but their transition into the surrounding 'empty' zones is usually quite smooth. An identification of 'sites' is clearly dependent upon an *a priori* notion of what a site should be in respect of the quantity of finds and their (typological) character and, of course, upon the scheme of fieldwork applied. Walking through landscapes with the aim of identifying 'sites', a lot of evidence stays behind and is not recorded, for instance, (1) individual artefacts and small scatters, the quantity of which is not sufficient 'to make a site', (2) large, low-density scatters that are empirically not recognizable, and (3) scatters where atypical artefacts and ecofacts prevail. Therefore, all site-oriented surveys are inevitably very selective, and miss not only much of the evidence on past non-residential activities (e.g. manuring scatters, production areas, communication zones), but also many regular residential areas that, for various reasons, may be represented by just a few artefacts on the surface.

Furthermore, 'sites' are never the integral spatial entities that they are often believed to be. 'Sites' mostly consist of remains from different periods of the past, none of which fully corresponds to the spatial extent of the surface artefact cluster as a whole. If whole 'sites' as spatial units are sampled by surface collection, a lot of information is lost on their inner spatial patterning. Any proper referencing of finds according to their chronological and/or other criteria can hardly be done in the field (before laboratory processing). What is usually lost, too, is the possibility of an accurate quantitative evaluation of the artefact density on the surface.

Considering these problems, it becomes clear that the site-oriented approach cannot work well within the scope of landscape archaeology and regional settlement studies. Instead, methods of 'siteless' survey, measuring surface artefact densities over larger segments of the landscape, should be articulated. We could call such methods *analytical* (Neustupný 1997a). Surface data are not approached as a set of empirically given spatial entities (sites), but their distribution is analysed in smaller reference units, which are to be covered systematically to provide quantitative and comparable values on surface artefact density. Although the application of such methods is certainly not new in the context of world archaeology (Foard 1978; Dunnell and Dancey 1983; Shennan 1985; Bintliff and Snodgrass 1985; 1988; Dunnell 1988; 1992; Gaffney and Tingle 1989), it has, however, never been tested on a large scale in the Czech Republic. Very little has also been known until recently of how such an analytical approach can work in the conditions of the Central European landscape and its archaeological record.

Most of the data used in this paper have resulted from the field activities of the Czech–British *Ancient Landscape Reconstruction in Bohemia* (ARLB) project, carried out by the Institute of Archaeology, Prague, and Sheffield University in 1991–95 (Beneš *et al.* 1992; Zvelebil *et al.* 1993; Kuna *et al.* 1993;

Figure 4.2   Ancient Landscape Reconstruction in Bohemia project, 'Central' transect. The distribution of survey polygons.

Kuna 1996; figs. 4.1–2). The ALRB was the first field project to develop and test some of the methods and techniques of analytical data collection, adapted to the specific character of the Bohemian landscape and its archaeological record. The aim of the project was to discover the long-term pattern of settlement change, complemented, of course, by some more detailed questions on settlement history at various spatial levels. The survey scheme developed in this project included (1) the systematic or randomized selection of survey polygons (fields) in the landscape; (2) the systematic coverage of the selected polygons by sampling units (sectors) of 100 × 100 m; and (3) the surveying of the units at an intensity of 10% (20 m between the survey traverses; cf. Kuna *et al.* 1993). This scheme was later adopted by the *Loděnice Iron Age Industrial Area* project (Venclová 1995; Neustupný and Venclová 1996) and by the recently inaugurated *Prehistoric Settlement of Bohemia* project (Department of Spatial Archaeology, Institute of Archaeology, Prague), combining surface artefact survey with an extensive use of aerial photography, geophysics, etc.

The cognitive limits of surface data

Surface artefact scatters are the remains of past activities that have been transformed by the dynamics of subsequent human behaviour and by the dynamics

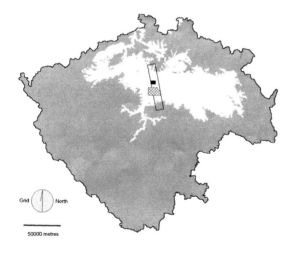

Figure 4.1   Bohemia. Ancient Landscape Reconstruction in Bohemia project, 'Central' transect (hollow oblong); Všetaty area (black); Vinoř-Creek project (crosshatched); altitudes above 300 m (grey).

of landscape development. Various aspects of formation processes introduce severe distortions into the original patterns of artefact distribution left behind at the loci where the artefacts were used and discarded. The major part of the evidence gained by the ALRB project consists of pottery fragments. Working with pottery data we are of course facing an even more complex set of problems than with some other sorts of finds (e.g. lithics), and we must be aware of even more secondary factors affecting the available evidence. This is also why Czech prehistoric archaeologists have so far approached surface data with considerable mistrust, and have pointed out many reasons why surface artefact collections bring information of limited explanatory value. The scepticism usually concentrates along the following lines:

1) The quantities of artefacts in the (surface) record depend on the variable cultural aspects of past settlement activities (Neustupný 1982; 1993b).
2) Artefacts may be displaced by erosion and recent ploughsoil transfers, and thus the reliability of surface scatters as indicators of past activity loci may be limited (Fridrich 1993; Vencl 1995).
3) Soil accumulation buries archaeological deposits below the level of ploughing, making surface survey ineffective (Vencl 1995; Neustupný 1982).
4) Surface artefact quantities and their visibility are dependent on variable ploughzone manipulation in the past and present, as well as on the actual conditions of the field surface at the moment of survey (Neustupný 1982; Vencl 1993).
5) Surface artefact assemblages usually have a low rate of chronological resolution (Vencl 1993).

In the following paragraphs I will not discuss all these topics, but only those that are particularly significant for the evaluation of data gained by analytical survey. Such a survey aims at obtaining a controlled sample of surface artefact densities over large parts of landscape. The intensity of survey is usually limited by the time schedule and personnel capacities of the project and, therefore, the extended spatial and quantitative information is usually paid for by the small numbers of artefacts obtained from individual sampling units. These data sets would, within a traditional approach, hardly be seen as something worth considering. The properties of these data can, however, be better understood within an appropriate model of formation processes.

Cultural and functional bias in pottery data

Pottery is generally a perishable material, with a relatively low resistance to the weathering and mechanical treatment to which it is exposed in the ploughzone and on the ground surface. Prehistoric pottery fragments, because of their poorly fired material, appear on the present-day surface only if, having originally been buried in deeper deposits, they have been ploughed out during the last few years or decades. Prehistoric pottery sherds that were left on the original surface long ago, or which were ploughed out further in the past, have most probably disappeared completely owing to their permanent rotation and fragmentarization in plough soil. Most of the variability in the frequency and density of surface pottery artefacts of individual prehistoric periods may thus be explained by the variable inclination of past populations to build subterranean features (houses, storage pits, etc.) at places of residence or other activity areas. The changing, culture-specific preference towards 'digging pits' might have had very little in common with the general function of activity areas, and more probably reflects some not yet completely understandable social or ideological aspects of settlement behaviour (Neustupný 1997b).

In any case, the number and character (size, depth) of deeper settlement features built at a certain activity locus during the time of its use are certainly the main factors affecting the quantity of pottery fragments available in the topsoil. This is definitely one of the reasons that certain prehistoric periods are quite 'invisible' in surface data (e.g. Corded Ware), although we know from other sources of evidence (cemeteries) that the settlement density must have been very high. No residential sites with deeper structures of the Corded Ware period have, for example, ever been excavated in the Czech Republic, and its pottery is also almost entirely missing in the surface data. In some other prehistoric periods we might face a similar situation, although in a more moderate way. For example, the frequency of Middle Eneolithic, Bell Beaker, Middle Bronze Age or Migration Period pottery in the surface data is also quite low. Again, this may be explained by particular cultural aspects of their residential behaviour, namely by the low numbers of deeper settlement features in their domestic areas (in some other periods the low density of data may be explained by other factors, too).

The information gained from prehistoric pottery scatters is, moreover, distorted not only from the cultural (regarding individual archaeological cultures and/or chronological periods) but also from the functional point of view. The reason for this is the same: prehistoric pottery does not generally survive in the ploughzone for a long time and therefore we cannot expect to identify activity areas where pottery was discarded on the surface, e.g. fields, some production areas, etc. This is why we can generally obtain very little information on prehistoric non-residential activities. For the same reason, we may, however, assume that prehistoric pottery is a relatively reliable spatial indicator of residential loci (since it most probably disappears completely before it can be moved by ploughing over a large distance).

This situation differs from the evidence obtained by more durable categories of artefacts and ecofacts. Hard-fired mediaeval pottery, for example, apparently survives quite well in the topsoil, which makes it possible to monitor manuring scatters around mediaeval villages. Some promising results have been also obtained by mapping the iron slag fragments and the sapropelite production refuse connected to Iron Age industrial activities (Neustupný and Venclová 1996; this volume). Unfortunately, the density of lithics in post-Mesolithic sites is generally rather low in most parts of the Czech Republic, which makes it difficult to complement the information obtained from pottery distribution by other types of data. (The number of prehistoric pottery fragments obtained during the ALRB project in its 'Central' transect, for example, reaches 25,000, which contrasts with less than 600 lithic items from all periods).

A quantitative model for surface pottery scatters

To evaluate the chances of surface survey identifying a certain type of residential area, it may be useful to articulate a quantitative model of how prehistoric residential loci can be manifested in surface scatters. Some calculations have been made (Kuna 1994; Salač 1995) on the density of pottery fragments in cultural debris in subsurface deposits of different periods. Values differ locally, but similar rates for whole residential areas are usually observed (the average density of pottery fragments in the fill of the settlement pits of various prehistoric periods mostly falls between 50–250 fragments per cubic metre). This might indicate that settlement behaviour in various prehistoric periods might have had roughly similar dynamics, at least in terms of the two main aspects responsible for total sherd numbers left behind at places of residence: intensity of pottery production and discard, and the duration of residential areas in time. We may assume that one prehistoric feature (settlement pit) covering an area of 2–15 sq m can provide (given a density of pottery fragments of 100 per cu m and that 20 cm of fill subsequently becomes ploughzone) some 40–300 items to the topsoil layer. Only 5–15% of this number (2–45 items) can appear on the surface, usually spread over an area of several hundred square metres. If the survey reaches an intensity of some 10% (walking in traverses 20 m apart), we may expect that such a cluster of finds will be cut by just one survey traverse, and will contribute only a few fragments, if any, to the resulting data. It is, therefore, not surprising that the find densities registered by systematic field traversing can often be very low, even at those locations where a prehistoric residential locus is present. We know that

prehistoric residential areas often consist of a small number of deeper features, sometimes spread over a large area; we cannot expect, therefore, to identify them as dense clusters of surface finds. This point, of course, must also be reversed: even low-density scatters or isolated finds should not be excluded from our data as evidence of secondary value ('off-site'), since they may represent a quite commonplace manifestation of residential areas for certain prehistoric periods.

The inconsistency of the ploughzone record

A critique of surface data usually points out that the densities of surface artefacts change over time at the same places (Vencl 1993; 1995). The change may happen in:

1) *the long term,* caused by the slow but gradual destruction of subsurface deposits by ploughing, the decay of artefacts or their covering by soil accumulation over decades or centuries
2) *the medium term,* changing artefact quantities due to crop rotation (cycles of 4–8 years) and the depth of soil treatment over the several years preceding the survey
3) *the short term,* changing the available artefact quantities over months, weeks or even days by their variable exposure to visual control, owing to the variable weathering of the surface, current soil treatment, uneven coverage by crops, etc.

The bias introduced by the long-term dynamics of the landscape surface can hardly be removed. It can only be *explained* if the history of the arable fields is considered. Study of land use over the most recent decades or centuries (the earliest available maps of Bohemia come from the end of the 18th century) is therefore recommended. In contrast to this factor, short-term change caused by weather factors and crop coverage at the moment of survey can be controlled and limited. In the ALRB project, for example, only those fields were surveyed whose surface had been exposed to some rainfall since the last ploughsoil treatment, and which were not densely covered by standing crop.

Changing surface artefact densities in the course of several years (the medium term), depending on agricultural cycles (crop rotation), represent the most serious danger for the consistency of surface data. The common model of ploughzone dynamics supposes that prehistoric pottery fragments appear on the field surface immediately after a deep ploughing event occurring once in 4–8 years (preceding, for example, maize, beet, potato), but that they may be heavily fragmented during the next season and decay quite soon. Some observations seem to confirm this model, but no accurate measurements have been done so far. Doing this, in fact, still remains an important task for future research. The data gained by the ALRB project did not bring clear evidence of the (medium-term) dynamics of surface scatters; they did, however, contribute some indirect observations reducing the above-mentioned model to a more moderate scale:

1) Surface data often produce large scatters crossing the boundaries of several land use units (fields), but no clear correlation between the type of crop and the density of pottery fragments or their average size is observable.
2) There were very few loci with extremely large pottery fragments being freshly ploughed out from deeper deposits. Most pottery scatters probably reflect a longer period of localization in the ploughzone: it seems, therefore, that pottery fragments do not decay as fast as assumed by the original model, and do remain in the ploughzone between the following episodes of the extremely deep ploughing (once in 4–8 years).
3) The model of changing ploughing depths during agricultural cycles need not be universal. It is usually applied to deeper soils; on shallower soils where the subsoil would be ploughed, the depth of ploughing does not change so dramatically. The speed of destruction of subsurface deposits primarily reflects, then, the rate of (colluvial, aeolian) erosion of the field surface, which makes the ploughing deeper even if the same techniques are applied. This would mean new artefacts coming to the ploughzone in a slower, permanent process, not as a result of sudden events occurring once in several years.
4) If the field surface is weathered and not covered by a dense crop (the only two conditions under which fields were walked in the ALRB project), there is little seasonal variation in average artefact densities over the year.

Chronological resolution

In general, prehistoric and mediaeval pottery assemblages in Central Europe are characterized by quite a good chronological resolution, since pottery of most periods displays distinctive attributes that can be recognized even on fragments. From the point of view of surface data, however, this statement is questionable; there are three factors making chronological information from surface pottery finds less complete and consistent than we would appreciate:

1) variable, often very low densities of surface pottery fragments;
2) variable average number and character of typological attributes available on pottery sherds of different periods;
3) variable fragmentation of pottery due to post-depositional processes.

First, the recognition of chronological components in surface pottery scatters is, basically, influenced by *the size of the collected sample* (Table 4.1). On average, about 12% of pottery fragments (calculated from data of the ALRB project) can be dated to individual prehistoric cultures or larger time segments (e.g. Neolithic, Bronze Age). Hence, a recognition of most chronological components within a survey sampling unit would theoretically require the collection of a sample where each of the components is represented by about 10 pottery fragments or more. In fact, such a sample is hardly achievable, since, as argued above, the surface density of finds from some periods may be quite low. Even the most intensive survey (impossible in large-scale projects) need not necessarily bring a sufficient number of finds for every component in each of the reference (sampling) units.

Second, *the character of chronologically distinctive attributes* is very heterogeneous. In some periods we may use specific technological attributes (some of the Neolithic, Iron Age or Mediaeval wares), other periods may be recognized by attributes deriving from vessel surface treatment and therefore still relatively abundant in an average pottery set (e.g. combing in the Final Bronze Age or Roman Period). Some periods, however, are distinctive neither by technology nor vessel surface treatment, and can be recognized

Number of pottery fragments per unit	Number of units (sectors)	Distribution of survey sectors according to the total artefact number and number of chronologically specific periods (%)						
		Number of periods						
		0	1	2	3	4	5	> 5
1	101	90.1	9.9	—	—	—	—	—
2–10	158	84.2	15.2	0.6	—	—	—	—
11–50	53	50.9	32.1	11.3	—	3.8	1.9	—
51–100	12	—	25.0	41.7	25.0	—	—	8.3
>100	14	—	14.3	—	21.4	28.6	21.4	14.3

Table 4.1 Ancient Landscape Reconstruction in Bohemia project, Všetaty area. Distribution of chronologically recognizable periods in survey units (squares 100 × 100 metres), in relation to total artefact (prehistoric pottery) numbers. Total number of units is 631.

only by fragments showing a particular vessel shape or a particular decoration detail (Eneolithic, Early and Middle Bronze Age). The probability of occurrence of a distinctive fragment is, therefore, rather different for individual periods. We can even presume that this lack of typological distinctiveness is another of the reasons that some prehistoric or Early Mediaeval phases are so scarcely represented in surface data (e.g. the Bylany culture of the earlier phase of the Hallstatt period; the Migration Period; Early Mediaeval 1 and 2: 6th–8th century AD). In contrast to the prehistoric cultures and the Mediaeval period before 800 AD, pottery wares from the 9th century onwards are much more recognizable, owing to their specific technologies and decoration.

Third, the chronological resolution of prehistoric pottery changes with the *rate of fragmentation*, depending on the timespan between the artefacts having been ploughed out from deeper contexts and the moment of survey. This may introduce further bias in the data, since distinctive attributes are variably sensitive to this effect (vessel shape 'disappears' very soon, surface decoration 'holds' longer, technology is often recognizable even on smaller and weathered fragments). As argued above, most of the pottery data collected during the ALRB survey reflect high fragmentation (hence a longer time of exposure in the ploughzone). This makes their general chronological resolution lower and can introduce bias in underrepresenting those periods of which the distinctive attributes are not technological but stylistic.

Data synthesis

Artefacts collected during an analytical survey usually display a wide spread over the landscape. The broad distribution of prehistoric pottery, for example, compared to a map of earlier finds and excavations, makes clear how fragmentary and biased our knowledge of the prehistoric settlement generally was and would remain without surface surveys, particularly those of the 'analytical' kind. Surprisingly but logically, data known from occasional finds and earlier excavations usually come from quite different parts of the landscape. They mostly reflect recent and subrecent building and mining activities inside villages, sand quarries and clay pits (areas not available for surface survey), but quite rarely has anything been registered in the surrounding open fields (surface artefact survey not being common until recently). This can be shown in some of the results of the ALRB project: from the Všetaty area, for example, there had been almost no information on the area corresponding to the project survey polygons, although it covers more than one-third (12.6 sq km; see Figure 4.3) of the territory, and more than half of the surveyed area includes prehistoric pottery scatters on the surface (Figure 4.4).

We may assume that most prehistoric and Early Mediaeval pottery scatters represent residential areas. As argued above, there is probably little reason to expect that manured fields or other types of non-residential activities would be identified through

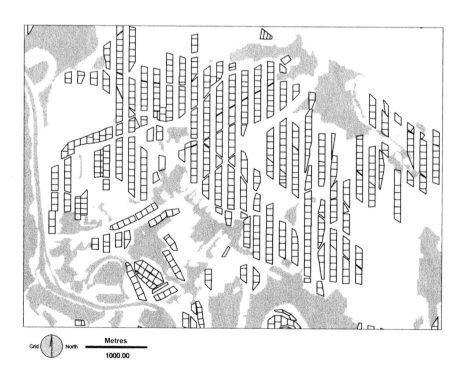

Figure 4.3   Ancient Landscape Reconstruction in Bohemia project, Všetaty area. The distribution of survey sectors (*c.* 1 ha sampling units). Built-up areas (villages) and woodland marked grey.

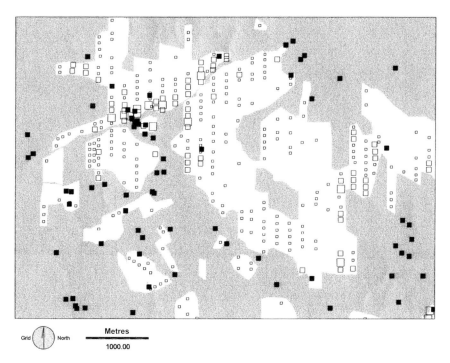

Figure 4.4   Ancient Landscape Reconstruction in Bohemia project, Všetaty area. Prehistoric pottery densities on the
surface (hollow squares); earlier excavations and finds (black solid squares). Surface find densities marked by
squares of three sizes (1–10; 11–100; >100 fragments per survey unit).

prehistoric or Early Mediaeval pottery (in contrast to
later periods, from the 13th century onwards).

Although quite rich, the collected pottery set is
clearly far from being a complete representation of
past settlement. Quite obviously, it is just a thin
sample of the records still surviving in the ground. We
do, however, believe that this sample still retains
structural information, the spatial and chronological
pattern of the settlement development in the long
term. An important piece of information, for example, may be the degree of continuity of residential
areas over the long-term development of farming
prehistory and the Early Mediaeval period. In the
Všetaty area, the overall distribution of the prehistoric pottery displays a definite pattern at first glance,
showing larger concentrations of finds at places that
may have served as permanent residential areas over
long periods of prehistory (Figure 4.4). The question
raised in our case study was whether these areas can
really be understood thus, and whether there was a
long-term continuity in the spatial behaviour of, and
cultural landscape formation by, past populations.

From the point of view of its description and classification the analytical data certainly present a difficult
task. To approach this data by purely empirical
methods, consisting of overlaying maps and their
intuitive interpretation, would most probably be useless, since the available surface evidence is not only
too complex but also still very fragmentary. On the
other hand, only this kind of data can be submitted to

quantitative and multivariate mathematical treatment. Therefore, a method of multivariate analysis
(principal component analysis [PCA]; vector synthesis
in terms of Neustupný 1993a) has been applied to the
data from the Všetaty area, exploring the distribution
of individual find categories and their combinations in
survey units. Within this part of the ALRB transect,
631 reference units (1 ha squares; Figure 4.3) were
surveyed; the available data set mostly consists of
pottery fragments. The composition of this data
according to individual periods is shown in Table 4.2.

In terms of its explanatory value, the data from the
ALRB project have variable reliability. The finds
numbers obtained for several pottery categories that
are usually available in larger quantities (the 'prehistoric' and Early Mediaeval 3–4 categories) can be
approached as true quantitative values reflecting
cumulative settlement intensity, modified of course
by depositional and post-depositional processes. With
some reservations, this sort of data may be used for
quantitative comparisons, and there is little need to
suspect that purely incidental factors may have caused
an absence of finds or a substantial quantitative mis-
representation of their surface density. In contrast to
these, the character of numbers of culture-specific
items is obviously different. The frequency of occurrence of distinctive fragments is generally very low
(with traverses 20 m apart), so that the probability of
missing them by survey is quite high. The *c.* 10%
intensity of survey does not, in this case, bring a

Find category	Date	Number of units (sectors)	Number of finds (pottery fragments)	Maximum per one unit
Neolithic	5500–4300 BC	19	27	4
Eneolithic	4300–2200 BC	14	22	7
Early–Middle Bronze Age	2200–1300 BC	13	18	3
Late Bronze Age	1300–1000 BC	22	41	9
Final Bronze Age	1000–750 BC	13	24	7
Bronze Age–Hallstatt		25	107	21
Hallstatt D–La Tène A	600–400 BC	11	19	4
La Tène B–La Tène D	400 BC–0	32	107	21
Roman Period	0–400 AD	24	153	36
Prehistory		327	4814	188
Early Mediaeval		35	95	17
Early Mediaeval 3	800–1000 AD	59	476	72
Early Mediaeval 4	1000–1200 AD	88	431	45
High Mediaeval	1200–1500 AD	426	2005	37

Table 4.2   Ancient Landscape Reconstruction in Bohemia project, Všetaty area. Frequency of occurrence of individual find categories (pottery fragments), total find numbers and maximum values per unit.

representative sample for each reference unit (as is the case for more abundant find categories), but only for the broader area of the survey. This is a result of the fact that the intensity of survey in this case not only affects the *number of fragments* obtained per unit, but also the *number of units* where any items of certain finds categories could be registered. Besides this, the occurrence as well as the recognition of culture-specific fragments is more affected by incidental factors (e.g. including the rate of fragmentation of pottery, local clustering of finds on the surface, etc.). For this reason, the observed numbers of pottery fragments were used as input data only in the case of the finds categories of prehistoric, Early Mediaeval 3 (9th–10th century AD), Early Mediaeval 4 (11th–12th century AD) and Early Mediaeval (unclassifiable more accurately, possible date between 6th and 12th century AD). In the case of other finds categories

(individual prehistoric cultures and periods), the find numbers were changed into binary variables (i.e. only presence/absence was considered). Besides this, an attempt to substitute to some extent for missing information was made by the 'filtering' of the data through a procedure that ascribed each unit the value of the average calculated from all the surrounding units within a radius of 200 m.

The results of the PCA are very clear. Five factors were considered, explaining about 80% of the variability in the data set (Table 4.3). At this level, we may infer that the residential areas of all the prehistoric periods were closely correlated in space (all of them show significant factor loadings within the dominant Factor 1). The only exception is the Neolithic, that is not significant in Factor 1 but forms the dominant element of Factor 2. The next three factors (3–5) were occupied by the Early Mediaeval

Find category	Factor 1	Factor 2	Factor 3	Factor 4	Factor 5
Neolithic	.21	**.90**	.13	.10	.05
Eneolithic	**.81**	−.09	.30	.14	−.02
Early–Middle Bronze Age	**.83**	.14	−.06	.04	.17
Late Bronze Age	**.83**	.33	−.08	−.14	.03
Final Bronze Age	**.70**	.24	−.10	−.11	.15
Bronze Age–Hallstatt	**.88**	.11	.21	.08	−.02
Hallstatt D–La Tène A	**.57**	.47	.04	−.03	−.20
La Tène B–La Tène D	**.78**	.28	.30	.10	−.02
Roman Period	**.67**	.56	.01	−.01	.00
Prehistory	**.66**	.37	.20	.26	.08
Early Mediaeval	.12	.12	**.94**	−.05	.01
Early Mediaeval 3	.04	.08	.04	**.97**	.03
Early Mediaeval 4	.09	.00	.01	.04	**.97**

Table 4.3   Landscape and Settlement (ALRBc) project, Všetaty area. Principal component analysis of the interpolated find densities. Rotated factor matrix. Bold fonts show the most significant factor loadings.

Figure 4.5   Ancient Landscape Reconstruction in Bohemia project, Všetaty area; principal component analysis. Factor 1 (post-Neolithic prehistory). Factor score values marked by squares of three sizes (0.5–1.0; 1.0–2.0; >2.0). Reconstructed streams according to historical maps, geological survey and aerial photography. Unsurveyed area marked grey.

periods, which correlate neither with the prehistoric finds nor with each other. The resulting factor scores were then displayed in space as GIS layers, identifying areas most significant for a certain find category or categories (Figures 4.5 and 4.6). This method — the combination of multivariate analysis with GIS — has been introduced to archaeology by E. Neustupný (cf. Neustupný 1996; Neustupný and Venclová 1996).

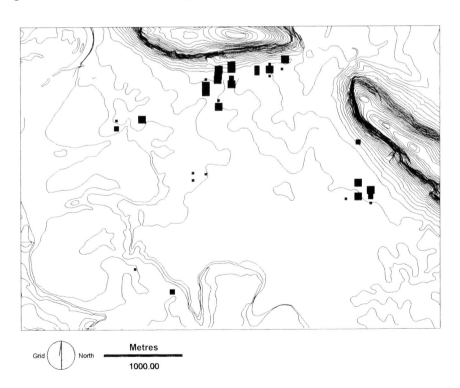

Figure 4.6   Ancient Landscape Reconstruction in Bohemia project, Všetaty area; principal component analysis. Factor 4 (Early Mediaeval 3). Factor score values marked by squares of three sizes (0.5–1.0; 1.0–2.0; >2.0). Contour lines at 2 metre intervals, total range of altitudes 160–238 m above sea level.

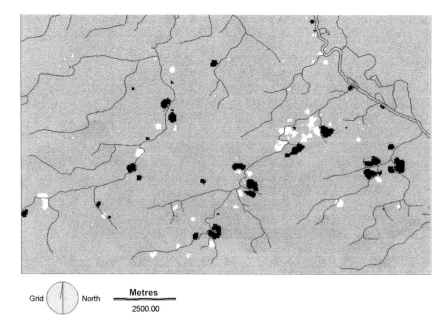

Figure 4.7 Vinoř-Creek project. Principal component analysis. Factor scores of Factor 1 (characteristic for Late Bronze Age and Final Bronze Age). White: factor scores lower than −0.5; grey: factor scores between −0.5 and 0.5; black: factor scores higher than 0.5.

Interesting results have also been obtained by applying a similar method, a PCA of settlement data interpolated in GIS, to the finds from the *Vinoř-Creek Survey project* (carried out by the author in the late 1980s: Kuna 1991a; 1997). These included surface materials collected by various strategies, earlier museum finds as well as data from several dozen smaller excavations. By means of PCA, a regular pattern was recognized in the landscape, displaying long-term residential areas as clusters of high factor scores. It can be assumed that each of these clusters corresponds to one prehistoric community and its residential area (Figure 4.7). In the next step of computer processing it was possible to ascribe

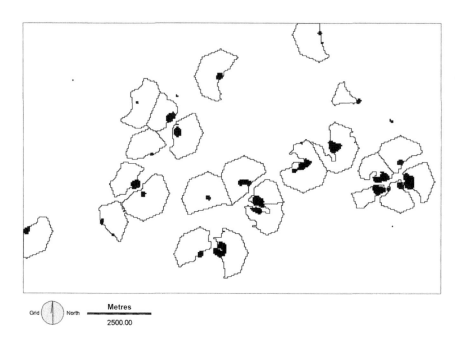

Figure 4.8 Vinoř-Creek project. Late to Final Bronze Age community areas calculated around high factor score clusters by GIS (one river side; Thiessen polygons; 1 km cost distance).

individual communities their 'community areas' as territories (1) lying on the same side of the stream as their residential areas; (2) belonging to the same 'Thiessen polygon'; and (3) being within the cost distance of 1 km from their residential areas (Figure 4.8). It is very interesting that the size range of such areas (100–150 ha) exactly fits into that which was predicted by D. Dreslerová (1995; Kuna 1997).

Interpretation: the continuity of landscape

The most interesting aspect of the results obtained by the PCA in the Všetaty area example is the surprising correlation of residential areas throughout all of pre-history, with the exception of the Neolithic. Looking at the map of scores for Factor 1 (Figure 4.7), we can see about 10 clusters of significant values that may be understood as continuous residential areas, cores for 'community areas' (Neustupný 1986; 1991). This continuity does not, of course, mean permanent settlements staying at the very same loci for centuries or millennia (this happens only after the 13th century AD), but rather a continuous oscillation of residential areas around some focal points in the landscape within a radius of some 200–400 m. The large spatial extent of these 'residential cores' is the reason why archaeology has so far so rarely considered a settlement continuity reaching over more than one or two successive prehistoric cultures. The earlier archaeological evidence was mainly based upon excavations that could of course never have been large enough to reveal the wider context of the excavated settlement segments, and it was difficult even to conceptualize such a research question.

The striking long-term continuity of prehistoric residential areas cannot be explained solely by invariable environmental factors (the location of streams, geology, soil quality, etc.). It is clear that these factors were considered by prehistoric populations, but the area of suitable environment was usually much larger than that which was really used for residential loci. Besides, the allocation of resources (as far as we can understand them) was not always strictly respected. It was not a lack of resources (at least in the area under study) that structured particular decisions in past settlement behaviour. It has been argued (Kuna 1998) that reasons for the continuity of residential areas should rather be sought in the formation and maintenance of cultural landscapes through time and, potentially, in the continuity of social meanings ascribed to certain places by people (Tilley 1994).

Residential areas represent the most visible (usually, from the surface, the only identifiable) component of community territories, but we have to assume and define other component types as well: burial areas,

ceremonial places, fields, pastures, woodland as a source of fodder, etc. Once the landscape had been structured by human behaviour these activity areas must have retained some stability, probably even higher than that of the residential places themselves. In some instances, this stability may have resulted from economic uses (fields, woodland), while in others it may have been supported by symbolic and ideological reasons (e.g. we know cases of prehistoric cemeteries used over the course of millennia). The stability of the landscape structure as a whole might also explain, in general terms, the continuity of residential areas. This brings us close to the notion of landscape as not just a set of natural elements, but as a social, cultural product of long-term diachronic stability.

Factors representing Early Mediaeval finds are also interesting for several reasons. First, it is quite surprising that the 'Early Mediaeval' (undatable more closely) finds correlate with neither the Early Mediaeval 3 (9th–10th century AD) nor the Early Mediaeval 4 (11th–12th century AD) categories. This opens the possibility of the interpretation of the 'Early Mediaeval' pottery not just as the accompanying material of the other two Early Mediaeval phases but, in fact, the residuum of a specific chronological phase that has not been recognized in the data. This could, most probably, only be the period of the 7th–8th century AD, the pottery of which has otherwise been recognized only in a very small number of cases (and which, therefore, has not been included in the factor analysis as a separate category). This hypothesis seems to be supported by the fact that the 'Early Mediaeval' finds often occur in the areas characteristic for Factor 1 that is, within the prehistoric residential areas. Supposing a general continuity of residential places, this may serve as another indication for their earlier date.

The separation of the following two Early Mediaeval phases as independent factors is obviously caused by the clustering of finds in areas that do not correlate with places intensively used before. The very fact that relatively short periods of time (roughly two hundred years each) produced large concentrations of finds and independent patterns in space may be understood as a hint that settlement behaviour in these periods was to some extent more coordinated. We may assume that there were some particular (political, social) factors that put an end to the 'longue durée' of the prehistoric settlement pattern and caused faster and more substantial shifts of residential areas. In the case of the Early Mediaeval 3 finds, we could guess that the nucleated residential area close to a gap in the mountain ridge reflects the presence of a central place appearing here in the 9th century AD, during the Early Mediaeval state formation process (Figure 4.6). In fact, the existence of an important Early Mediaeval hill-fort was supposed on the top of the neighbouring hill (Přívory). Although this has recently been questioned (Sláma 1988: 69), the density and spatial extent of surface finds (supported by the PCA results) at the bottom of the hill may nevertheless bear witness to some more important role for the residential area in this place.

Conclusions: future tasks

General issues discussed in this paper may be summarized as several points relevant, in my opinion, to the further development and effective future application of surface artefact survey in the Czech Republic:

(1) The coming decades will probably bring *more problems for surface artefact surveys* when compared to the last 40–50 years. The intensification of agricultural production and the application of new techniques in the 1950s and 1960s brought the destruction of many subsurface archaeological contexts. Due to this, many new artefact scatters appeared on the surface, disappearing again or becoming less visible during the following years and decades (this applies mainly to prehistoric pottery; mediaeval ceramics, and lithics are more durable). Czech archaeology has almost entirely missed this chance, being preoccupied with other research topics. Current less destructive agricultural techniques, the (re)privatization of land and changes in land use will probably make the access to, and the identification of, the surface archaeological record even more difficult. Despite this, the surface record still contains a *vast information potential for archaeology*, that can hardly be substituted for by any other kind of data.

(2) Czech archaeology should adopt the basic principles of *analytical surface survey,* which does not consider the archaeological record as a set of isolated 'sites' but rather as a 'continuous data surface' spread over the landscape. It should also adopt surface survey as a legitimate *means of primary data collection*, not only as a technique of archaeological 'reconnaissance' bringing useful but still preliminary information that in itself should not be used for theoretical interpretations.

(3) There is an urgent need to experiment with various *sampling schemes* in defining an optimal field survey methodology adapted to the character of surface data in the Czech Republic. The efficacy of sampling schemes is of vital importance for the validity of the resulting data. Besides that, a specific limitation of surface artefact surveys in the Czech Republic results from the prevailing *climatic conditions*. The optimal season for fieldwalking does not usually last longer than some 4–5 weeks per year, and it may be difficult to predict exactly when it will come. It is, therefore, necessary not only to collect *repre-sentative* data but also to do so within a reasonable *timespan*. The size of the units, their distribution in space and the intensity of survey inside them are interconnected variables affecting data validity and increasing/decreasing the time and labour costs of the survey; therefore, these aspects must be thoroughly considered and articulated in an explicit model. The training of specialists, students and non-professional archaeologists in this type of fieldwork may also be crucial.

(4) Methods should be developed enabling us to cope with the specific character of data obtained by analytical methods. Amongst other problems, analytical survey usually brings a very thin sample of the available surface record that is itself nothing more than a very small fraction of the past reality. Therefore, it is often necessary to consider even *very small numbers of artefacts* as archaeologically significant. As I have tried to show in this paper, small finds numbers result from the cultural aspects of past settlement behaviour and ploughzone taphonomy. We should get rid of the traditional notion that one or two pottery fragments per hectare mean nothing and should be sorted out as 'background noise' (Kuna 1994; Neustupný and Venclová 1996). In many cases, surface artefact counts cannot be much higher even if the survey intensity increases.

(5) It is believed that even low find numbers may contain the pattern of past behaviour and important archaeological information. The necessity of working with small numbers should lead to the definition of appropriate *methods of mathematical data synthesis,* since the intuitive evaluation of this data is clearly impossible. One such method, using a multivariate procedure in combination with GIS, has recently been developed by E. Neustupný (1996).

(6) The most important issue for the future is a better understanding of the *processes transforming archaeological remains in the ploughzone*. There are many observations on this topic but still very few exact measurements: we do not know exactly how long prehistoric pottery survives in the topsoil, how much its quantity changes over time, etc. Without learning this, surface artefact survey can hardly improve its 'image' among other types of archaeological fieldwork and appear as a mature technique able to provide relevant data on the past.

References

Beneš, J., and Koutecký, D.
1987   Die Erforschung der Mikroregion Lomský potok. Probleme und Perspektiven. In E. Černá, (ed.), *Archaeologische Rettungstätigkeit in den Braunkohlengebieten und die Problematik der siedlungsgeschichtlichen Forschung*, 31–38. Prag: Archäologisches Institut.

Beneš, J., M. Kuna, L. Peške, and M. Zvelebil
1992 Rekonstrukce staré kulturní krajiny v severní části Čech: československo-britský projekt po první sezóně výzkumu (Reconstruction of the ancient cultural landscape in North Bohemia). *Archeologické rozhledy* 46: 337–42.

Bintliff, J., and A. Snodgrass
1985 The Cambridge/Bradford Boeotian expedition: the first four years. *Journal of Field Archaeology* 12: 123–61.
1988 Off-site pottery distributions: a regional and interregional perspective. *Current Anthropology* 29: 506–13.

Břicháček, P., and L. Košnar
1987 Mikroregion dolní Cidliny v době římské a nové nálezy importované keramiky (Die Mikroregion an der unteren Cidlina in der römischen Kaiserzeit und neue Funde importierter Keramik). *Archeologické rozhledy* 39: 557–69.

Buchvaldek, M.
1965 Terénní archeologický průzkum. Základní pokyny pro činnost dopisujících spolupracovníků. *Zprávy ČSSA* 7(2–3) 14–19.

Charvát, P.
1981 Povrchový průzkum Litomyšlska a Vysokomýtska. *Výzkumy v Čechách 1975 — Supplementum*, 73–77.

Dreslerová, D.
1995 A socio-economic model of a prehistoric microregion. In M. Kuna and N. Venclová (eds.), *Whither Archaeology? Papers in honour of E. Neustupný*, 145–60. Praha: Institute of Archaeology.

Dunnell, R.C.
1988 Low-density archaeological records from plowed surfaces: some preliminary considerations. *American Archaeology (Issues in archaeological surface survey)* 7-1: 29–38.
1992 The notion site. In J. Rossignol and L.A. Wandsnider (eds.), *Space, Time and Archaeological Landscapes*, 21–42. New York: Plenum Press.

Dunnell, R.C., and W.S. Dancey
1983 The siteless survey: a regional scale data collection strategy. In M.B. Schiffer (ed.), *Advances in Archaeological Method and Theory* 6: 267–87.

Fencl, V.
1975 Archeologické nálezy z terénních průzkumů v okolí Velvar. *Výzkumy v Čechách 1973*: 247–55.

Foard, G.
1978 Systematic fieldwalking and the investigation of Saxon settlement in Northamptonshire. *World Archaeology* 9(3): 357–74.

Fridrich, J.
1993 Comments about the problem of spatial archaeology. *Památky archeologické* 84(1): 155–56.

Fröhlich, J., and J. Michálek
1989 Archeologický průzkum území dolního toku Blanice. *Archeologické výzkumy v jižních Čechách* 6: 7–41.

Frolík, J., and J. Sigl
1995 *Chrudimsko v raném středověku. Vývoj osídlení a jeho proměny* (Chrudim region [East Bohemia] in the Early Middle Ages. Development of settlement and related structural changes). Hradec Králové: Muzeum Východních Čech.

Gaffney, V.L., and M. Tingle
1989 *The Maddle Farm project: an integrated survey of prehistoric and Roman landscapes on the Berkshire Downs*. BAR British Series 200. Oxford: British Archaeological Reports.

Hammer, F.
1964 Archeologické lokality z nejbližšího okolí katastru obce Mutějovice, okr. Rakovník. *Výzkumy v Čechách* 2: 129–35.
1966 Archeologické lokality na katastru obce Mutějovice, okres Rakovník. *Výzkumy v Čechách* 4: 75–92.

Jahnkuhn, H.
1955 Methoden und Probleme siedlungsarchäologischer Forschung. *Archaeologica geographica* 4: 73–84.
1976 *Archäologie und Geschichte. Vorträge und Aufsätze, Band 1. Beiträge zur siedlungsarchäologischer Forschung*. Berlin: W. de Gruyter.

Klápště, J.
1985 Archeologiczne Zdjęcie Polski — polský projekt a česká skutečnost. *Archeologické rozhledy* 37: 347–49.

Klápště, J., and J. Žemlička
1979 Studium dějin osídlení v Čechách a jeho další perspektivy. *Československý časopis historický* 27(6), 884–905.

Knor, A.
1954 Nálezy na chmelnicích v severozápadních Čechách (Trouvailles sur les houblonnières de la Bohême Nord-occidentale). *Památky archeologické* 45: 281–300.

Kolbinger, D.
1995 Pokračování systematického povrchového průzkumu východní části kroměřížského okresu v roce 1994. *Informační zpravodaj ČAS, pobočka pro Moravu a Slezsko*, prosinec 1995: 24–35.

Kudrnáč, J.
1961 Rekonstrukce přirozené krajiny v okolí zkoumaných hradišť a osad (Die Rekonstruktion der natürlichen Landschaft in der Umgebung der durchforschten Burgstätten und Gemeinden). *Památky archeologické* 52: 609–15.

Kuna, M.
1991a The structuring of prehistoric landscape. *Antiquity* 65(247): 332–47.
1991b Návrh systému evidence archeologických nalezišť (A descriptive system for a database of archaeological sites). *Archeologické fórum* 2: 25–48.
1994 Archeologický průzkum povrchovými sběry (Archaeological survey by surface collection). *Zprávy České společnosti archeologické — Supplément*: 23.
1996 GIS v archeologickém výzkumu regionu: vývoj pravěké sídelní oblasti středních Čech (GIS in regional archaeological research: development of prehistoric settlement patterns in central Bohemia). *Archeologické rozhledy* 48: 580–604.
1997 Geografický informační systém a výzkum pravěké sídelní struktury (GIS and prehistoric settlement patterns). In J. Macháček (ed.), *Počítačová podpora v archeologii*, 173–94. Brno: Masarykova univerzita.
1998 Keramika, povrchový sběr a kontinuita pravěké krajiny (Ceramics, surface survey and the continuity of prehistoric landscapes). *Archeologické rozhledy* 50: 192–223.

Kuna, M., M. Zvelebil, P.J. Foster and D. Dreslerová
1993 Field survey and landscape archaeology research design: methodology of a regional field survey in Bohemia. *Památky archeologické* 84(1): 110–30.

Meduna, P., and E. Černá
1991 Settlement structure of the Early Middle Ages in northwest Bohemia: investigations of the Pětipsy basin area. *Antiquity* 65: 388–95.

Neustupný, E.
1965   Hrob z Tušimic a některé problémy kultur se
        šňůrovou keramikou (The grave of Tušimice and
        some problems of the Corded Ware cultures).
        Památky archeologické 56: 392–456.
1982   Optimalizace výzkumu archeologického regionu
        (Die Optimalisierung der Erforschung archäolo-
        gischer Region). In *Metodologické problémy česko-
        slovenské archeologie: Konference čs. archeologů k
        60. výročí založení Archeologického ústavu ČSAV*
        (Liblice, 13–14.11.1979): 178–81. Praha: Institute
        of Archaeology.
1984   Archeologická prospekce s využitím pravděpo-
        dobnostních metod (Prospecting by means of
        probabilistic methods). In Nové prospekční
        metody v archeologii. *Výzkumy v Čechách —
        Supplementum:* 105–30.
1986   Sídelní areály pravěkých zemědělců (Settlement
        areas of prehistoric farmers). *Památky archeolo-
        gické* 77: 226–34.
1991   Community areas of prehistoric farmers in
        Bohemia. *Antiquity* 65(247): 326–31.
1993a   *Archaeological Method.* Cambridge: Cambridge
        University Press.
1993b   Some field walking theory. *Památky archeologické*
        84(1) 150–52.
1996   Polygons in archaeology. *Památky archeologické*
        87: 112–36.
1997a   Analytické a syntetické sběry. Unpublished
        manuscript.
1997b   Šňůrová sídliště, kulturní normy a symboly
        (Settlement sites of the Corded Ware groups,
        cultural norms and symbols). *Archeologické rozh-
        ledy* 49: 304–21.
Neustupný, E., and N. Venclová
1996   Využití prostoru v laténu: region Loděnice (Geb-
        rauch des Raumes in der Latènezeit: die Region
        Loděnice). *Archeologické rozhledy* 48: 615–42.
Pleiner, R., and A. Rybová (eds.) 1978, *Pravěké dějiny
        Čech.* Praha: Academia.
Pleinerová, I., and J. Muška
1981   Terénní průzkum lokality Březno a jeho okolí.
        Výzkumy v Čechách 1975 — Supplementum: 5–9.
Rada, I.
1987   Die siedlungsarchäologische Erforschung in
        Becken von Pětipsy. In E. Černá (ed.), *Archaeolo-
        gische Rettungstätigkeit in den Braunkohlengebieten
        und die Problematik der siedlungs-geschichtlichen
        Forschung,* 39–42. Praha: Institute of Archaeology.
Salač, V.
1995   The density of archaeological finds in settlement
        features of the La Tène period. In M. Kuna and
        N. Venclová (eds.), *Whither Archaeology? Papers
        in honour of E. Neustupný,* 264–76. Prague:
        Institute of Archaeology.
Sedláček, F.
1967a   Archeologická naleziště na katastru obce Mšec,
        okres Rakovník. *Výzkumy v Čechách* 5: 71–77.
1967b   Archeologická naleziště v nejbližším okolí Mšece:
        Srbeč, Trtice, Mšecké Žehrovice (okres Rakov-
        ník). *Výzkumy v Čechách* 5: 79–81.

Shennan, S.
1985   *Experiments in the Collection and Analysis of
        Archaeological Survey Data: The East Hampshire
        Survey.* Sheffield: Department of Archaeology,
        University of Sheffield.
Sláma, J.
1988   Střední Čechy v raném středověku. III. Arche-
        ologie o počátcích českého státu (Central Bohe-
        mia in the Early Middle Ages, III. Archaeology
        and the beginnings of the Přemysl-dynasty state).
        Praehistorica 15. Prague: Univerzita Karlova.
Smetánka, Z.
1970   Zur Methodik von Feldforschungen an mittelal-
        terlichen Ortswüstungen (Auszug aus einem
        Referat). *Časopis Moravského muzea — Acta
        Musei Moraviae* 55: 63–70.
Smetánka, Z., and J. Škabrada, J.
1975   Třebonín na Čáslavsku v raném středověku —
        povrchový průzkum (Die Gemeinde Třebonín in
        der Čáslaver Gegend im frühen Mittelalter).
        Archeologické rozhledy 27: 72–85.
Smrž, Z.
1986   Entwicklung und Struktur der Besiedlung in
        der Mikroregion des Baches Lužický potok. In
        E. Černá (ed.), *Archaeologische Rettungstätigkeit
        in den Braunkohlengebieten und die Problematik
        der siedlungs-geschichtlichen Forschung,* 17–30.
        Prague: Institute of Archaeology.
1987   Vývoj a struktura osídlení v mikroregionu
        Lužického potoka na Kadaňsku (The develop-
        ment and structure of settlement in the micro-
        region of the stream Lužický potok in the area of
        Kadaň). *Archeologické rozhledy* 39: 601–21.
Soudský, B.
1966   *Bylany. Osada nejstarších zemědělců z mladší doby
        kamenné* (Bylany. Station des premiers agricul-
        teurs de l'âge de la pierre polie). Praha: Academia.
Thrane, H.
1989   Siedlungsarchäologische Untersuchungen in
        Dänemark mit besonderer Berücksichtigung von
        Funen. *Prähistorische Zeitschrift* 64(1): 5–47.
Tilley, C.
1994   *A Phenomenology of Landscape: Places, Paths and
        Monuments.* Oxford: Berg.
Vencl, S.
1968   Povrchový sběr jako technika archeologického
        průzkumu. *Muzejní a vlastivědná práce* 6(2): 96–99.
1993   Comments about the ALRB — Landscape and
        Settlement projects. *Památky archeologické* 84(1):
        154–55.
1995   K otázce věrohodnosti svědectví povrchových
        souborů (Surface survey and the reliability of its
        results). *Archeologické rozhledy* 47: 11–57.
Venclová, N.
1995   Specializovaná výroba: teorie a modely (Specia-
        lized production: theories and models). *Arche-
        ologické rozhledy* 47: 541–64.
Zvelebil, M., J. Beneš, and M. Kuna
1993   Ancient Landscape Reconstruction in Northern
        Bohemia — Landscape and Settlement Pro-
        gramme. *Památky archeologické* 84: 93–95.

5. Reflections on the Future for Surface Lithic Artefact Study in England

John Schofield

Summary

Since 1988 the Monuments Protection Programme (MPP) of English Heritage has been undertaking a thorough review of England's archaeological resource, with a view to making informed decisions on its future management. For many types of site this review has been relatively straightforward, for example, where there has been a long history of research, and where classes of monument are well known and understood as a result. For some types, however, we have had to start more or less from scratch, commissioning or undertaking research to establish the basic framework: how many sites of a certain type are there? where are they located? and so on. Surface lithic material falls between the two extremes: there has been much work but little attempt at synthesis. As a result the data have succeeded in putting dots on maps, but our understanding of settlement history and patterns of use have progressed little. A national survey of existing records aims to change all that, making the data useful and useable at a macro-scale for the first time.

Background

> Much depends on the status of flint scatters, still a major factor in the record. Lacking the context of time depth and of function as they invariably do, they remain no more than a simple record of presence/absence although attempts have been made to elevate them to the level of true settlement pattern. (Kinnes 1994: 95)

In England surface artefact study has an established role within the field of settlement archaeology, and over the years this role has engaged approaches to site delimitation, intra-site patterning and regional survey. It has been used in research projects, management surveys and, more recently, as one of a suite of techniques used in evaluation work undertaken in the context of development control. The potential of surface artefact study in all of these situations, and in each approach, I consider to be beyond question. Whether that potential has yet been fully realized,

however, is quite another matter, and it is that which forms the subject of this short chapter. On this occasion I will confine myself to the question of regional survey, the scale of analysis for which surface lithic artefact study is arguably best suited and most used, at least in England.

Regional survey, examining the distribution of material remains at a macro-scale, now has a significant role in studying the past. And this is equally the case whether our concern is with human interaction with nature, adapting to the physical environment, or is more to do with how the landscape was perceived by those who moved around it — Bradley's notion of imagining the landscape (1997), and the studies of phenomenology presented by Tilley for Cranborne Chase (1994) and Bodmin Moor (1996), for instance. Such regional studies allow us to view patterns of exploitation and use at a broad scale, embracing a variety of landscape types and viewing change over long periods. The information is necessarily coarse, in terms of spatial and temporal control, contrasting with the fine-grained, precise detail that can generally be gleaned from excavation. And when settlement sites *per se* appear only exceptionally in a region's archaeological record, as is the case for much of mainland Britain, and certainly central southern England, during the Neolithic and early Bronze Age periods, such coarse-grained data become invaluable for providing information on settlement, both as an aspect of human activity and its physical manifestation.

This paper will address the situation in England, using the example of Wessex during the later prehistoric period. It will include brief discussion of the history of surface artefact study, as well as describing current approaches and recent exemplars; however, the main theme will be future directions, based on the premise that to look forward we first need to take stock of our present position, evaluating current records as a means to define priorities for future research. This is the approach currently being taken by English Heritage, and an outline of the methodology and preliminary results of its national review of surface lithic material is presented. This paper supplements various others published recently that describe various aspects of this survey: these cover its background, rationale and methodology (Schofield 1994a; Schofield and Humble 1995); theoretical context (Schofield 1995); and the value of surface lithic scatters within the wider framework of conservation philosophy (Lisk, Schofield and Humble, 1998).

Historical context

The surface collection of artefacts has been practised in England for over 200 years. Naturally there has been regional bias (towards such areas as Wessex, for instance), as well as significant variations in what at the time merited collection and retention, what was deposited with museums and what was later discarded (as 'rubbish' presumably). Much of this early work represented collection for its own sake, though into the 20th century this became increasingly rare, with a growing realization that such artefacts had an academic value, providing significant information about past settlement and land use strategies. In Wessex, and Hampshire specifically, the collections of Chris Draper are of particular note. Draper's fieldwork dates largely to between the 1930s and 1950s, and was concentrated mainly on the area of the Meon valley in south-east Hampshire. His work was extensive but unsystematic, covering the full range of topography and including areas of chalk downland no longer under cultivation. Unlike many of his contempories Draper published his results in a number of brief reports, some of which present data (e.g. Draper 1953), and others synthesis (e.g. Draper 1955). The extent to which the results of such early collection can be incorporated into current thinking on settlement and land use strategies is reflected in various studies (e.g. Gardiner 1987; Schofield 1994b).

However, surface artefact study only really became established as a worthwhile archaeological endeavour in the 1960s, a time when perceptions of and approaches to the past were changing. At this time ideas central to the so-called New Archaeology, such as the spatial continuity of human behaviour, came to redefine the objectives and methods of surface artefact study in England. The work of Glynn Isaac (1981) and Robert Foley (1981), with their models of what was collectively termed off-site archaeology, Stephen Shennan's work, in particular the use of quantitative methods, culminating in his East Hampshire Survey (1985), and surveys in the United States, such as the Cannon Reservoir Human Ecology Project (O'Brien *et al.* 1982), all in the decade 1975–85, contributed to give surface artefact study credibility in the broader context of landscape studies.

Research design and problem orientation also influenced the nature and organization of survey in England at this time, providing a much-needed focus. In 1978 Tim Schadla Hall and Stephen Shennan defined a sampling approach to archaeological survey in Wessex (Schadla Hall and Shennan 1978), and, 20 years after, it is interesting to see how closely that has been followed: many of the potential survey areas identified by Schadla Hall and Shennan were investigated and the results are now published, some representing benchmarks in terms of presentation and interpretation. As we have seen, Shennan's East Hampshire survey was one of the first to apply quantitative methods in defining significance in the patterning of surface material, while work around Stonehenge (Richards 1990) and on Cranborne Chase (Barrett *et al.* 1991) provided a context for some of our best-known ceremonial monuments, identifying areas of contemporary settlement and industrial activity. Additional work focused deliberately on those areas of the region where funerary and ceremonial monuments were a less common occurrence (Schofield 1991a; Light *et al.* 1994). Table 5.1 contains a summary of the major surface artefact surveys (or surveys involving a significant surface collection component) undertaken in Wessex over the last 20 years, giving a summary of dates and objectives. It is perhaps important to stress that few (if any) other areas in England have seen this intensity of collection, undertaken by a variety of individuals and organizations, and this has created its own difficulty, namely the comparison and compatibility of results. This is discussed further below. A more coordinated approach has been taken in the Fenland Survey (Hall and Coles 1994), a region of England that comes close to (and may exceed) Wessex in terms of the total area subjected to surface artefact collection.

The current emphasis in England towards research undertaken within well thought-out and agreed frameworks maintains that focus first defined in the mid 1970s. At a national level, this approach, and the philosophy that underpins it, is set out in English Heritage's *Frameworks for our Past* document (Olivier 1996), while an example of local and regional frameworks is an edited volume *Archaeology in Hampshire* (Hinton and Hughes 1996), containing chapters that cover the main periods and in which authors review the current state of knowledge and suggest priorities for future research. In the chapter on early farming communities, for instance, Julie Gardiner (1996) suggests as priorities the closer comparative examination of surface artefact scatters, especially from non-chalk areas; the identification and analysis of assemblages from the coastal plain and two major river valleys; and transect collection strategies around major groups of contemporary burial monuments. Similarly, Peter Woodward, in the Dorset Ridgeway Survey, noted how 'much more clearly needs to be done to provide a firm basis for future comparative research of a material that is sometimes ill-defined in terms of chronology, cultural

Reference	Dates of survey	Survey area	Aims and objectives (extracts/summary only)
Shennan 1985	1977–78	East Hampshire	To gather reliable information as a basis for systematic study; to collect data in such a way that it could be used to try and investigate its significance in relation to such things as collection bias, surface conditions, etc.
Richards 1978	1976–77	Berkshire Downs	'To collate known archaeology and attempt to assess destructive processes taking place and the archaeological potential of the area.'
Richards 1985	1980–86	Stonehenge environs	'To identify the prehistoric settlements in the Stonehenge region, and to establish their state of preservation ...'
Ford 1987	1984–86	East Berkshire	'... improve the quantity and quality of data in the SMR by undertaking an extensive survey programme.' 'The survey attempted to collect a consistent set of data to test whether human activity through time varied according to base geology.'
Gaffney and Tingle 1989	1981–82	Maddle Farm (Berkshire Downs)	'(1) To determine differential functions of areas within settlement complexes; (2) to determine the nature of land-use associated with contemporary settlements; (3) to establish and interpret any existing settlement hierarchy, including the definition of social/tenurial relationships; (4) to locate contemporary settlement distribution.' (Although framed with Roman settlement in mind, the authors go on to state that they apply equally to earlier and later periods.)
Schofield 1991a	1984–86	Upper Meon valley	'... to investigate the nature and intensity of prehistoric occupation in a part of southern England not generally considered a "core area" (and) to define patterns of land-use and settlement within the upper reaches of a small chalkland river valley.'
Light *et al.* 1995	1979–86	Middle Avon valley	To assess an area of Wessex that had received little previous attention. Specifically, to consider whether the influence of the neighbouring core areas of the upper Avon valley and Cranborne Chase had 'rubbed-off' on this region.
Sharples 1991	1985–86	Maiden Castle environs	To collect data to enable an historical analysis of the landscape surrounding Maiden Castle and to clarify the significance of the hilltop occupation. The systematic collection of artefacts to produce detailed distribution maps was a central component of the survey.
Barrett *et al.* 1991; Arnold *et al.* 1988	1977–84	Cranborne Chase	'... what had started as an investigation of landscape history ... extended into a study of social change. As this happened ... attention turned to the role of the more spectacular field monuments in Cranborne Chase [and] their relationship to the contemporary pattern of settlement.'
Boismier 1994	1984–89	Hampshire County Council land-holdings	To provide information on the County Council's land-holdings. Also to explore local patterning in the distribution of artefact classes across the region, and the potential to partition the landscape into site and non-site areas; assess area classification on the basis of site size, artefact density and artefact diversity; explore the relative temporal ordering of areas.
Lobb and Rose 1996	1976–77; 1982–87; 1988–89	Lower Kennet valley	'... to assess the archaeological resource of an area that has seen considerable development pressure within the past two decades and to respond to the consequent threat to it.'
Woodward 1991	1977–84	South Dorset Ridgeway	'(1) To explore the archaeological record exposed by modern arable cultivation and to identify the impact of cultivation on preservation; (2) to develop fieldwork techniques and excavation sampling strategies which could identify prehistoric habitation sites and settlements; (3) to explore the relationship between burial monuments, settlements and land-use.'

Table 5.1   Dates and objectives of the major surface artefact surveys in Wessex.

identity and function' (1991: 126). But, as we shall see, to establish research frameworks requires us first to know the resource sufficiently; we need to have a feel for what is known and what is not known before the questions we really want answers to can be targeted.

Right here, right now

To give an idea of the scale of, and potential in, the existing records, a stocktake of the national resource of surface lithic material is being undertaken (Schofield 1994a; 1995; Schofield and Humble 1995). This has, for the first time, established an estimate of what data already exist, and what types of information can be gleaned from them. For instance, we now know that in 4 of the 50 or so English counties (as defined prior to recent local government reorganization), a total of 3300 surface lithic scatters and 1750 stray finds are recorded. If this is multiplied up (given that this study was designed to be representative), over 41,000 surface lithic scatters and nearly 22,000 stray finds might be recorded in England. In fact the true figure will be higher still, as many of these scatters are multiperiod (in the sense that they fall within palimpsests and are thus recorded more than once). There are those who consider surface lithic scatters 'rubbish data', but surely (and in terms of quantity alone) a database of that size must contain *some* useful and useable information? Or maybe the very size and the complexity of surface data, combined with the difficulties of interpretation referred to below, present problems that render them potentially useful but not necessarily useable?

Academic demands

Any attempt to define the value of surface scatters and to establish future directions, whether in terms of methodology or research priorities, must take account of the demands being made and likely to be made of the data, and in England we are beginning to get some idea of what those demands might be. Recently researchers have started using previously collected data in developing general ideas of the ways landscape was perceived and used in prehistory. Examples include Julie Gardiner's (1984) work, which explored land use in prehistoric Wessex through studying surface artefacts from a time before systematic collection. Chris Tilley, in his book *Phenomenology of Landscape* (1994) explored surface lithic material in relation to topography, assessing the significance of landscape features in small-scale societies. And Chris Gosden (1994) used surface material, alongside other archaeological remains, in developing ideas on temporal frameworks. Thus, as well as requiring the overview from which research frameworks can derive, we need to be aware of the growing demands now being put on the data, and this can be seen both in terms of the number of enquiries, and the sophistication of the questions being asked.

Of course, there are certain complicating circumstances that need to be taken into account when interpreting surface scatters, and these are well known. For instance, only a fraction of the plough-soil assemblage will be visible on the field surface, and this will vary according to artefact size and shape; the weather and surface conditions during collection; the extent to which erosion and deposition have affected soil depth; and the length and frequency

of occupation episodes represented — an industrial site used intensively on one or two occasions can generate significantly more lithic material than a persistent place, repeatedly occupied over centuries; collection bias has also been recorded as significant in determining recovery in some surveys. These are only some of the variables — they are well documented and need no further mention here (cf. various papers in Schofield 1991b and Wandsnider and Camilli 1992, which address these issues). The point is the extent to which some consider these factors — when taken together — effectively to devalue surface material and render it useless for analytical purposes. I disagree. If the question is whether a habitation site can be defined spatially and temporally, as if by excavation, the answer in most cases is of course that it cannot. But this isn't a question of rubbish data; rather one of an inappropriate question being asked. Equally, two or three excavations within a region make little headway in defining patterns of occupation and broad chronological trends; surface survey can do this. Horses for courses, as the saying goes.

So it is then a question of approach and understanding the limitations in the data we have. Most recent surveys do appreciate these limitations and generally present their results in terms of the cumulative and repetitive nature of human activity. But presenting results where conclusions refer to general trends rather than the specifics of site placement, size and form creates another problem (and one already identified above in quotes from Gardiner and Woodward): one of compatibility — comparing objectively the results of one survey with those of another. Compatibility is a significant consideration. As Chisholm once put it: '[I]t is important to be able to compare one pattern with another, since it is only in this way that explanations can be offered for the distributions that one is specially interested in' (1975: 63). I have attempted to address this, comparing the results of the middle Avon with the upper Meon valley (see below) and East Hampshire surveys (Schofield 1987: 276–84); Timothy Earle of University of California Los Angeles attempted something similar, though Wessex-wide, some years ago (this was inconclusive and is unpublished), but those have been amongst the very few attempts and little progress has been made. As a consequence, the regional dimension of surface lithic data is rarely included in synthetic volumes on British prehistory — Julian Thomas's *Rethinking the Neolithic* (1991) is one of the few exceptions.

Curatorial demands' and a response

Thus there is a clear academic demand currently being made of the data. A word now on the demands of curators (in a sense this is a false distinction:

curatorial demands *are* academic, and should certainly be framed within a wider academic context, a point I return to below). Since 1990 when a Government advice note was published in England, stating that archaeology was a material consideration in the planning process and that there should be a presumption of *in situ* preservation for nationally important archaeological remains (DoE 1990), curators have been concerned with the concept of national importance as it applies to the various types of archaeological site. Much of this work has been coordinated through the Monuments Protection Programme of English Heritage, whose role is to advise the British Government on such matters in England. (For general information about archaeological resource management in England, see Hunter and Ralston 1993; for details about the Monuments Protection Programme, see Startin 1995. Together these provide the context for the project described here.) At the present time, however, surface-collected data are not well enough understood for such definitions of importance to be drawn. Knowing how many scatters there are and where they are is a first step, but we need to go further if any such selection is to have credibility, and if research objectives and resources are to be targeted effectively.

These demands together provide the justification for an English Heritage project, undertaken jointly by the Central Archaeology Service and the Monuments Protection Programme, with a view to consolidating the existing database, assessing the data and establishing what we know and don't know about

surface lithic material nationally. The Surface Lithic Scatter Sites and Stray Finds Project was set up in 1993, initially with a pilot study covering four English counties (Schofield and Humble 1995). This study is now complete, and to present this briefly I will concentrate first on the methodology, and second on a few of the results.

First, methodology (detailed in Schofield 1994a). From existing sources, which included principally the Sites and Monuments Records (held by each of the counties), all stray finds and scatters were recorded. Details of location and date were recorded for each entry, as were various other factors, including the degree of survival, integrity (whether the scatters are known to be discrete), documentation (in terms of further work undertaken), association with other types of contemporary monument, the size of the scatter (number of artefacts) and the interpretation attributed to it by the collector or author. The source of the data was also identified and subsequently analysed as a means to assess the frequency with which surface scatters are entered on to Sites and Monuments Records, and how many are recorded only as museum accessions. The data for each of the criteria were recorded in such a way that they could be integrated with a GIS, and subsequently investigated using statistical methods.

At the simplest level, the results tell us how many scatters and stray finds there are, and where they are (Figure 5.1). We also now have some general information on the degree of precision with which details of provenance have been recorded. This ranges from

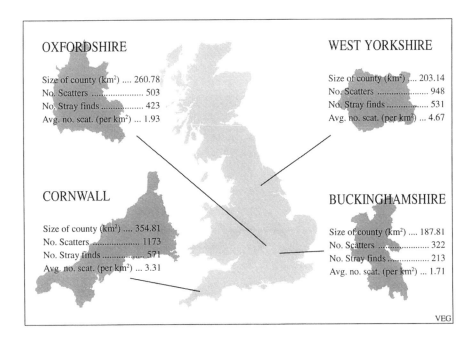

Figure 5.1 Descriptive statistics from the pilot study. Surface Lithic Project, English Heritage.

10 Ordnance Survey coordinates, which, if correct, denote location to within 1 m, through two-figure coordinates (accurate to 10 km), to no more than a parish name. It is encouraging to note that some 90% of scatters recorded in the pilot study can be located to an accuracy of less than 100 m and 36% to less than 10 m (Humble and Schofield 1996). So records of provenance are generally good, contrary to popular belief.

Statistical analysis of the results (undertaken for us by Kris Lockyer and Stephen Shennan) provides more detailed insight. For example, in terms of assemblage size, there is little difference between the counties, despite there being significant differences in the history of collection and the nature of the archae-

ology. In general, small and very small assemblages (<49 artefacts) make up between 78 and 85% of all scatters. The periods represented by scatters are of interest in terms of the relative numbers of single and multi-period scatters (Figure 5.2; single period scatters represent 43% of the total, for example), while division by period shows how few Palaeolithic scatters there are, how many scatters are not dated, and how similar the figures for Mesolithic and Neolithic material appear to be. The figure for scatters containing only Neolithic material has particular relevance in England, where only 50 locations have produced Neolithic buildings, only a selection of which may represent 'houses' in the true sense (Darvill 1996). The location and form of the living places of Neolithic communities in England have therefore not surprisingly been on the research agenda for some time. Some 340 Neolithic-only scatters were identified in the four counties studied, which multiplies up to a possible 4000 nationally. So, in understanding Neolithic settlement, these data must be significant and we should do all we can to read them effectively.

In terms of integrity (the extent to which scatters are discrete or spread), results show, first, that most scatters are spread and do not have recognizable boundaries, and, second, that the data show significant regional variations (Figure 5.3a). Although it is tempting to suggest that these figures may relate to regional variations in the nature of prehistoric settlement, the level of systematic work undertaken may provide a more likely explanation.

The results also show the extent to which non-systematic survey has predominated (Figure 5.3b), the fact that most recorded scatters still survive in some form (Figure 5.3c), and that in 94% of cases no additional work has been undertaken on them (geophysical survey or excavation for instance) (Figure 5.3d). Generally speaking there exists a clear correlation between the level of survey undertaken and the degree to which scatters can be dated. Unsystematic survey (amounting in most cases to an unstructured search of field surfaces) tends to produce dateable material in the form of diagnostic artefacts; certainly this was true in the early years of surface collection. Systematic survey, on the other hand, is usually concerned with all humanly struck lithic material, a tiny proportion of which may be diagnostic. Function (defined here as the nature of activity attributed to the scatter by the collector or author) will inevitably be of interest to those exploring regional trends in land use and settlement. Alarmingly, 93% of all scatters have no known function. So, of the 1173 scatters in one of the counties (Cornwall), only around 60 could be attributed a function, even defined in such broad terms as settlement, industrial or ceremonial. Some have gone so far as to say, 'If that is the case, what is the point?'

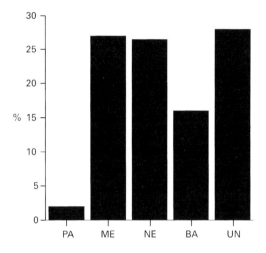

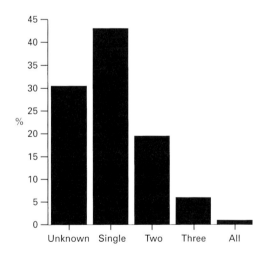

Figure 5.2   Scatters by period. Top, bar chart of scatters arranged by period. PA = Palaeolithic; ME = Mesolithic; NE = Neolithic; BA = Bronze Age; UN = undated. Bottom, bar chart of multiperiod scatters in which the numbers refer to the number of these periods represented in multiperiod scatters.

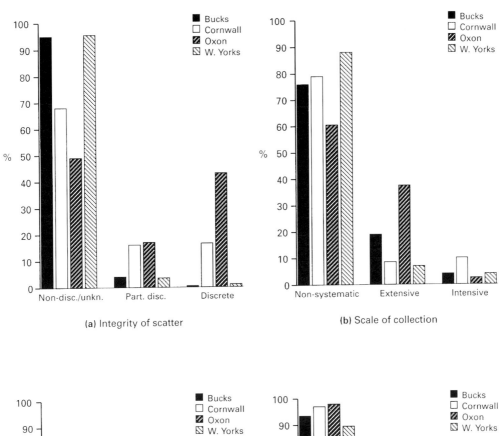

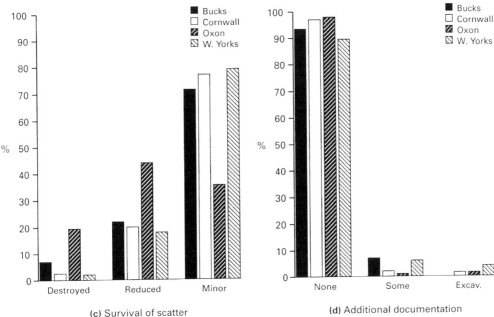

Figure 5.3   Bar charts showing some of the results of the pilot study, arranged by county. Non-disc./unkn. = non-discrete/ unknown; part. disc. = partially discrete; minor = minor alterations only to original scale of scatter; excav. = excavated.

Future directions

Although not inclusive, what follows is a list of research themes that I believe could usefully follow from the stocktake described above. Research frameworks (and beyond them, research agenda — statements of intent) require consensus, however, and this is something the profession is moving towards. Volumes such as this are valuable in presenting options and ideas, though in this case any pan-European agenda will require implementation at national or regional levels (in other words, the administrative frameworks within which professional archaeology is currently organized). At the national level in England, English Heritage are currently developing research frameworks and agenda through an extensive programme of consultation. This initiative (and its successors) will do much to determine the future course of surface artefact study in England. However, in the meantime, a few thoughts of my own will give some idea of the options.

First, methodology. Thought should be given to the effectiveness of our methodology. Perhaps there is scope for something more radical than that currently practised. Certainly, we should constantly be appraising and where necessary reviewing methodology, and specifically the way data are collected and results presented. Also, there needs to be some standardization in the way records of surface artefact study are made available to the curators responsible for maintaining and evaluating them in the context of development control. Procedures are available for estimating and comparing the adequacy of survey strategies (Sundstrom 1993) and this should be helpful in this regard, but some standardization would also prove invaluable to curators and those engaged in the study of existing results, for instance, making compatibility and regional comparisons more straightforward. As an example of this, I compared results from the upper Meon and middle Avon valleys, in south-east and west Hampshire respectively, and some interesting conclusions were drawn. Three variables were compared, reflecting the intensity of activity (measured by artefact frequency and density); the nature of activity (assemblage composition, including the core reduction sequence, retouch and technological indices); and chronology (measured in terms of a 'ubiquity index': the percentage of collection units within the survey area that produced items diagnostic of a single period). The results of this comparative exercise suggested clear contrasts in the land use strategies between the two areas. The upper Meon valley was a landscape considered physically well-suited for occupation and exploitation in the Mesolithic and early Neolithic periods, and this is reflected in the predominance of finds of this period made both prior to and during the survey (compared in Schofield 1994b). The scarcity of finds beyond that period, combined with the reverse pattern in the middle Avon valley, which lies close to two of the core areas — the 'monument zones' — of later prehistoric Wessex, appears to indicate emphasis rather towards the exploitation of emergent landscapes in the more political climate of the middle and later Neolithic. This is a point seemingly reflected in the manpower required for the construction of sites like Avebury, Stonehenge and, before them, the Dorset cursus, all generally interpreted as ceremonial centres serving a dispersed population (Shennan 1986: 147). So, it would seem from this that different parts of Wessex were favoured in the earlier and later Neo-

lithic periods, with the change of emphasis resulting from political developments that manifested themselves also in the emergence of monument-building. Further and more sophisticated comparative studies could enable us to move on from this simple statement to develop a more refined model of the emergence of social complexity.

Thus there is clearly potential for more studies into intra-regional comparison of results from surface artefact study. But I do not believe a standard methodology would be appropriate or indeed necessary for easing comparability, and hence a consideration of the various methodologies adopted in surface artefact study in England are not considered relevant here. However, some model format does need to be addressed, to which the presentation of results should conform for inclusion on archaeological record systems. There are, for instance, certain key facts, defined at the lowest common denominator, which any systematic survey should provide. Artefact density per unit area for instance (corrected according to surface area investigated), and the relative proportions of different artefact classes (e.g. number of primary flakes per core) would be helpful in compiling regional syntheses. As the example above shows, such comparison could then be straightforward both within and between survey areas, irrespective of the collection strategy adopted, enabling us to 'compare one pattern with another', the thing — Chisholm suggests — that interests us most (1975: 63). It is encouraging to see this level of compatibility now being addressed in other countries (e.g. Bintliff 1997).

There are of course other questions, one of which concerns the value of undertaking systematic survey alone, when we know from the pilot study that diagnostic artefacts result more from extensive non-systematic work, and that function can rarely be attributed to the results of intensive systematic survey. Should survey therefore include both systematic intensive collection, and the rapid unsystematic trawling of the countryside for the diagnostic items which give those results chronological context? And should we make more use of ploughsoil excavation, as has been suggested by Steinberg (1996)? And finally, with the presumption of *in situ* preservation, and present concerns about the storage capacity of museums, is there a need for surface collection at all? Why not the *recording* of surface scatters, just as we do with sites represented by earthwork remains or crop-marks, lifting the artefacts to record details of type, size, etc. before returning them to the surface from whence they came?

Second is the question of integrating surface lithic data with the results of recent and current initiatives that come under the umbrella of 'character mapping'. English Heritage, in conjunction with other national and regional organizations, has undertaken much work recently on character mapping, looking to define regions (including historic regions) by their distinctive qualities. One example of this is the work being undertaken by Brian Roberts, mapping the settlement zones of 19th-century England within which the character of mediaeval — and perhaps earlier — settlement will be distinguishable (Roberts *et al.* 1996). Other projects, one the work of English Heritage and the Countryside Commission, and another by English Nature, have attempted to view the fabric of the countryside at a national level; comparable work has also been undertaken regionally, in Cornwall and Avon, for instance (cf. various papers in Fairclough *et al.* 1999). Although the principal objective of this work is to promote a common national framework for conservation decisions of all kinds within the larger context of planning and agricultural policies, there is also an academic sense in which the results will be of considerable benefit. Integrating it with archaeological data, available in a consistent way and at a national scale, could yield some explanations at least for the distribution and form of material culture.

Also at this national level, and returning to points made earlier, surface lithic data need to be integrated more effectively with other archaeological evidence, as well as with data from palaeoenvironmental studies. To give an example of the former, a recent paper by Harding (1995) explored regionality in the English Neolithic through the form of causewayed enclosures, cursus and henges. These regional differences appear to persist throughout the Neolithic and may reflect fundamental distinctions between the make-up of separate political communities. If so, might we not expect also to see contrasts in some aspect of the settlement pattern and its material culture? Additionally, we need to appreciate better the parameters within which the material record was formed. It seems to be accepted now, for instance, that attempting to find lithic scatters representing settled communities in the Neolithic of southern England is a pointless exercise; communities at this time were mobile and material remains will be more akin to the diffuse and low-density spreads of nomadic groups than we had previously supposed (cf. chapters by Bintliff and Kuna, this volume). On the subject of integrating palaeoenvironmental and surface artefact studies, this is now becoming a feature of regional survey reports, as seen, for example, in the modelling of landscape development and modification around Stonehenge based on molluscan studies (Richards 1990: 254–57), and its incorporation into a general model of prehistoric land use embracing also surface lithic data (Richards 1990: 263–80).

Finally, the national study outlined above will highlight where the gaps in our knowledge lie, and, set

against national and regional research frameworks and agenda, these can be assessed and prioritized accordingly. Julie Gardiner has identified some possibilities in Hampshire (above) that could well form the basis for future work, just as Schadla Hall and Shennan's (1978) 'frameworks' did for Wessex 20 years ago. The difference now is the need for a national context, and the availability of results from many studies that could usefully be integrated with surface data to achieve a more holistic view of this significant part of our historic resource.

Acknowledgement

I am grateful to John Bintliff for inviting my participation at EAA in Riga, 1996, and to my employer, English Heritage, for enabling me to attend. Sincere thanks also to Jon Humble for his considerable assistance in managing the lithic scatters project; to those who undertook the study in the four counties; and to Kris Lockyer and Stephen Shennan, then of Southampton University, now of University College London, for undertaking the statistical analysis and appraisal of the pilot study, some results from which are included here. Finally, it should be noted that some of the content of this paper also appears in a separate publication (Schofield; 1997).

References

Arnold, J., M. Green, B. Lewis and R. Bradley
  1988 The Mesolithic of Cranborne Chase. *Proceedings of the Dorset Natural History and Archaeological Society* 110: 117–25.
Barrett, J. C., R. Bradley and M. Green
  1991 *Landscape, Monuments and Society: The Prehistory of Cranborne Chase*. Cambridge: Cambridge University Press.
Bintliff, J.
  1997 Regional survey, demography, and the rise of complex societies in the ancient Aegean. Core-periphery, neo-Malthusian, and other interpretive models. *Journal of Field Archaeology* 24: 1–38.
Boismier, W.A.
  1994 *The Evolution of the Hampshire Landscape: Archaeological Resources on County Council Owned Farm and Recreation Land*. Archaeological Report No. 2. Winchester: Hampshire County Planning Department.
Bradley, R.
  1997 Working the land: imagining the landscape. *Archaeological Dialogues* 4(1): 39–48.
Chisholm, M.
  1975 *Human Geography: Evolution or Revolution?* London: Pelican.
Darvill, T.
  1996 Neolithic buildings in England, Wales and the Isle of Man. In T. Darvill and J. Thomas (eds.), *Neolithic Houses in Northwest Europe and Beyond*, 77–111. Oxbow Monographs 57. Oxford: Oxbow.
DoE
  1990 *Archaeology and Planning*. Planning Policy Guidance 16. London: HMSO.

Draper, C.
1953 Mesolithic sites in south Hampshire. *Archaeological News Letter* 4: 60–61.
1955 Mesolithic and Neolithic distribution in southeast Hampshire. *Archaeological News Letter* 5: 199–200.

Fairclough, G., G. Lambrick and A. McNab
1999 *Yesterday's Landscape, Tomorrow's World: The English Heritage Historic Landscape Project.* London: English Heritage.

Foley, R.
1981 A model of regional archaeological structure. *Proceedings of the Prehistoric Society* 47: 1–18.

Ford, S.
1987 *East Berkshire Archaeological Survey.* Occasional Paper No. 1. Berkshire: Department of Highways and Planning, Berkshire County Council.

Gaffney, V., and M. Tingle
1989 *The Maddle Farm Project: An Integrated Survey of Prehistoric and Roman Landscapes on the Berkshire Downs.* BAR (British Series) 200. Oxford: British Archaeological Reports.

Gardiner, J. P.
1984 Lithic distributions and settlement patterns in central-southern England. In R. Bradley and J. Gardiner (eds.), *Neolithic Studies*, 15–40. BAR (British Series) 133. Oxford: British Archaeological Reports.
1987 Tales of the unexpected: approaches to the assessment and interpretation of museum flint collections. In A.G. Brown and M.R. Edmonds (eds.), *Lithic Analysis in Later British Prehistory: Some Problems and Approaches*, 49–65. BAR (British Series) 162. Oxford: British Archaeological Reports.
1996 Early farming communities in Hampshire. In D. Hinton and M. Hughes (eds.), *Archaeology in Hampshire: A Framework for the Future*, 6–12. Winchester: Hampshire County Council.

Gosden, C.
1994 *Social Being and Time.* Oxford: Basil Blackwell.

Hall, R., and J. Coles
1994 *Fenland Survey: An Essay in Landscape and Persistence.* English Heritage Reports (New Series) 1. London: English Heritage.

Harding, J.
1995 Social histories and regional perspectives in the Neolithic of lowland England. *Proceedings of the Prehistoric Society* 61: 117–36.

Hinton, D., and M. Hughes (eds.)
1996 *Archaeology in Hampshire: A Framework for the Future.* Winchester: Hampshire County Council.

Humble, J., and J. Schofield
1996 The Lithic Scatters Project: where does it come from and where is it going? *CAS News* 5: 7.

Hunter, J., and I. Ralston (eds.)
1993 *Archaeological Resource Management in the UK: An Introduction.* Stroud: Alan Sutton/IFA.

Isaac, G.
1981 Stone-age visiting cards: approaches to the study of early land use patterns. In I. Hodder, G. Isaac and N. Hammond (eds.), *Pattern of the Past: Studies in Honour of David Clarke*, 131–55. Cambridge: Cambridge University Press.

Kinnes, I.
1994 The Neolithic in Britain. In B. Vyner (ed.), *Building on the Past*, 90–102. London: Society of Antiquaries.

Light, A., A.J. Schofield and S.J. Shennan
1994 The Middle Avon Valley Survey: a study in settlement history. *Proceedings of the Hampshire Field Club and Archaeological Society* 50: 43–101.

Lisk, S., A.J. Schofield and J. Humble
1998 Lithic scatters after PPG16 — local and national perspectives. *Lithics* 19: 24–32.

Lobb, S., and P. Rose
1996 *Archaeological Survey of the Lower Kennet Valley, Berkshire.* Wessex Archaeology Report No. 9.

O'Brien, M.J., R.E. Warren and D.E. Lewarch
1982 *The Cannon Reservoir Human Ecology Project: An Archaeological Study of Cultural Adaptations in the Southern Prairie Peninsula.* New York: Academic Press.

Olivier, A.
1996 *Frameworks for our Past: A Review of Research Frameworks, Strategies and Perceptions.* London: English Heritage.

Richards, J.
1978 *The Archaeology of the Berkshire Downs: An Introductory Study.* Berkshire Archaeological Committee Publication No. 3. Reading: Berkshire Archaeological Committee.
1990 *The Stonehenge Environs Project.* London: English Heritage.

Roberts, B.K., S. Wrathmell and D. Stocker
1996 Rural settlement in England: an English Heritage mapping project. Praha: *Památky archeologické — Supplementum* 5: 72–79.

Schadla Hall, R.T., and S.J. Shennan
1978 Some suggestions for a sampling approach to archaeological survey in Wessex. In J.F. Cherry, C. Gamble and S.J. Shennan (eds.), *Sampling in Contemporary British Archaeology*, 49–65. BAR (British Series) 50. Oxford: British Archaeological Reports.

Schofield, A.J.
1987 Putting lithics to the test: non-site analysis and the neolithic settlement of southern England. *Oxford Journal of Archaeology* 6(3): 269–86.
1991a Lithic distributions in the upper Meon valley: behavioural response and human adaptation on the Hampshire chalklands. *Proceedings of the Prehistoric Society* 57: 159–78.
1991b *Interpreting Artefact Scatters: Contributions to Ploughzone Archaeology.* Oxbow Monographs 4. Oxford: Oxbow.
1994a Looking back with regret; looking forward with optimism: making more of surface lithic scatters. In N. Ashton and A. David (eds.), *Stories in Stone*, 90–98. Lithic Studies Society Occasional Paper 4. London: Lithic Studies Society.
1994b The changing face of 'landscape' in field archaeology: an example from the upper Meon valley. In M. Hughes (ed.), *The Evolution of the Hampshire Landscape: The Meon Valley*, 1–12. Hampshire County Council Archaeology Report No. 1. Winchester: Hampshire County Council
1995 Settlement mobility and la longue durée: towards a context for surface lithic material. In A.J. Schofield (ed.), *Lithics in Context: Suggestions for the Future Direction of Lithic Studies*, 105–13. Lithic Studies Society Occasional Paper 5. London: Lithic Studies Society.
1997 Staring through stereograms: the future for surface artefact survey? *Proceedings of the Latvian Academy of Sciences*, 48–53.

Schofield, A.J., and J. Humble
  1995   Order out of chaos: making sense of surface stone age finds. *Conservation Bulletin* 25: 9–11.

Sharples, N.M.
  1991   *Maiden Castle: Excavations and Field Survey, 1985–6*. English Heritage Archaeological Report No. 19. London: English Heritage.

Shennan, S.J.
  1985   *Experiments in the Collection and Analysis of Archaeological Survey Data: The East Hampshire Survey*. Sheffield: Department of Archaeology, University of Sheffield.
  1986   Interaction and change in 3rd millennium bc western and central Europe. In A.C. Renfrew and J.F. Cherry (eds.), *Peer Polity Interaction and Socio-Political Change*, 137–48. Cambridge: Cambridge University Press.

Startin, B.
  1995   The Monuments Protection Programme: protecting what, how and for whom? In M.A. Cooper, A. Firth, J. Carman and D. Wheatley (eds.), *Managing Archaeology*, 137–45. London: Routledge.

Steinberg, J.
  1996   Ploughzone sampling in Denmark: isolating and interpreting site signatures from disturbed contexts. *Antiquity* 70(268): 368–92.

Sundstrom, L.
  1993   A simple mathematical procedure for estimating the adequacy of site survey strategies. *Journal of Field Archaeology* 20: 91–96.

Thomas, J.
  1991   *Rethinking the Neolithic*. Cambridge: Cambridge University Press.

Tilley, C.
  1994   *A Phenomenology of Landscape: Places, Paths and Monuments*. Oxford: Berg.
  1996   The powers of rocks: topography and monument construction on Bodmin Moor. *World Archaeology* 28(2): 161–76.

Wandsnider, L., and E.L. Camilli
  1992   The character of surface archaeological deposits and its influence on survey accuracy. *Journal of Field Archaeology* 19: 169–88.

Woodward, P.J.
  1991   *The South Dorset Ridgeway: Survey and Excavations, 1977–84*. Dorset Natural History and Archaeological Society Monograph Series No. 8. Dorchester: Dorset Natural History and Archaeological Society'.

6. Territoire et Peuplement en France, de l'Age du Fer au Moyen Age. L'Archéologie Spatiale à la Croisée des Chemins

Claude Raynaud

Summary

Le développement des recherches sur l'occupation du sol constitue depuis une décennie l'une des orientations majeures de la recherche française. Suscitées conjointement par l'élaboration de la Carte Archéologique du Ministère de la Culture et par plusieurs projects de recherche, ces opérations englobent de nombreuses zones pilote, en Bretagne, dans le Bassin Parisien, le Centre et dans le Midi rhodanien. Ces travaux suscitent un renouvellement de la problématique et des méthodes, les prospections traditionnelles se trouvant remplacées par une approche plus systématique, associant étroitement recherches sur le terrain, carto- et photo-interprétation, études textuelles et analyses statistiques. Plusieurs milliers d'établissements inédits ont ainsi été inventoriés, bouleversant en bien des points les interprétations traditionnelles. S'ouvrent des perspectives nouvelles pour cerner l'organisation, la hiérarchie at la dynamique du peuplement. Une telle démarche s'est accompagnée d'une révision des concepts. En particulier, la notion de site archéologique perd sa centralité avec l'irruption de l'espace et du territoire, tandis que les concepts de système spatial et de réseau hiérarchisé s'imposent progressivement.

Ces recherches ont bénéficié d'un soutien de la Communauté Européenne et portant sur l'impact humain dans la vallée du Rhône (Van der Leeuw 1995). Les moyens mis en oeuvre dans ce programme ont permis d'insérer les données archéologiques dans un Système d'Information Géographique (GIS) et d'accéder à des outils de mesure et d'analyse statistique ainsi qu'à un croisement très poussé des données.

Introduction

Depuis les années 1980, l'archéologie française comble progressivement son retard dans le domaine de l'occupation du sol et du peuplement. La dernière décennie a vu s'accentuer l'effort de mise à jour de la carte archéologique, tant par la quantité des informations que par la qualité des données enregistrées. En Bretagne, dans le Val de Loire, en Languedoc et en Provence (Figure 6.1), plusieurs programmes régionaux ont contribué ces dernières années à la formation d'équipes pluri-disciplinaires, en cours d'intégration dans l'organisation administrative de la recherche. Le mouvement s'est développé au sein des trois principaux acteurs de l'archéologie nationale: le Centre National de la Recherche Scientifique avec le financement de plusieurs Actions Thématiques Programmées sur les questions spatiales, la Direction du Patrimoine du Ministère de la Culture avec le projet ambitieux de la Carte Archéologique, l'Université avec le recrutement de jeunes enseignants orientés vers cette problématique et avec la multiplication des thèses sur le sujet. Longtemps archaïques, les méthodes d'identification et d'inventaire

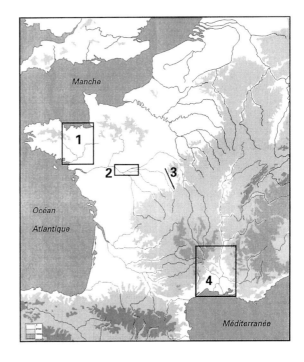

Figure 6.1   Localisation des zones mentionnées. Haute-Bretagne (1), Val de Loire (2), Berry (3), Zone ArchæoMedes: TGV-Méditerranée, Languedoc oriental et Provence occidentale (4).

Figure 6.1   Location of zones mentioned. Upper Brittany (1), Loire Valley (2), Berry (3), ArchaeoMedes zone: TGV-Mediterranean, Eastern Languedoc and Western Provence (4).

des sites ont rapidement progressé, en empruntant aux équipes anglo-saxonnes et en les adaptant aux contraintes physiques régionales, la collecte en carroyage, l'échantillonnage rigoureux du mobilier, les confrontations surface/fouille. Cette croissance a conduit à l'affirmation (trop rapide?) de cette jeune discipline, donnant à ses praticiens une (trop?) grande confiance dans la valeur de leurs résultats.

Collecte des données et élaboration de la carte archéologique

L'avancement des modes de collecte des données doit autant aux recherches 'programmées' (CNRS, Université) qu'aux travaux d'archéologie préventive (Sous-Direction de l'Archéologie, Ministère de la Culture). On soulignera comme un signe encourageant la convergence progressive des deux types d'approche, trop longtemps opposées en termes d'efficacité et de 'scientificité'. Souhaitons que s'estompe encore ce clivage entre une archéologie programmée (ou fondamentale) de haut niveau mais d'une lenteur désespérante, et une archéologie de sauvetage (ou appliquée), riche, rapide mais déconnectée de l'analyse historique. Dans un cadre comme dans l'autre, les équipes ont soumis techniques et méthode à de rigoureux contrôles. Un chantier

comme la construction de l'autoroute A71, dans la région Centre, a donné lieu à une expérimentation poussée et à l'évaluation de l'efficacité respective des différents types de prospection (Ferdière et Rialland 1994; 1995). On a pu ainsi mesurer le degré d'adéquation de chaque technique de prospection aux différents types de milieu. La prospection pédestre et visuelle a confirmé son efficacité en milieu labouré, moyennant le respect d'un certain nombre de paramètres garantissant la représentativité des 'images de surface'. Dans ce domaine, la prospection 'en ligne' demeure le plus large dénominateur commun entre les travaux entrepris (Figure 6.2) et il y a fort à parier qu'une telle approche, pratiquée avec rigueur et systématique, demeure une pratique dominante. Qu'on en juge d'après les chiffres: parmi les 237 sites découverts sur le tracé de l'autoroute A71, près de 60% ont été localisés selon cette technique. Parallèlement, la collecte en carroyage se répand, soit comme technique exclusive, soit comme approche d'appoint en aval de la prospection en ligne, dont elle vient pallier les insuffisances. On aurait bien tort d'opposer les deux techniques qui au contraire se complètent parfaitement, la première permettant d'embrasser relativement vite des échantillons territoriaux significatifs au plan historique, la seconde intervenant pour affiner la typologie et la hiérarchie des établissements à occupation longue et/ou complexe. La prospection

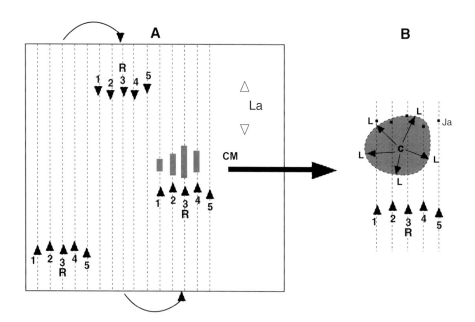

Figure 6.2   La prospection en ligne des milieux labourés. A : cheminement d'une équipe et découverte d'un établissement (R = responsable de l'équipe; CM = concentration de mobilier; La = sens du labour). B: délimitation d'une concentration de mobilier (C = centre théorique de la concentration; L = limites estimées; Ja = jalon marquant l'arrêt momentané de la progression; Ferdière et Rialland 1994: 33).

Figure 6.2   Line-walking in ploughed terrain. A: trajectory of a team and discovery of a site (R = team-leader; CM = surface scatter; La = direction of ploughing). B: delimitation of an artefact focus (C = theoretical centre of the site; L = estimated boundaries; Ja = ranging pole marking the temporary stop point in fieldwalking; Ferdière and Rialland 1994: 33).

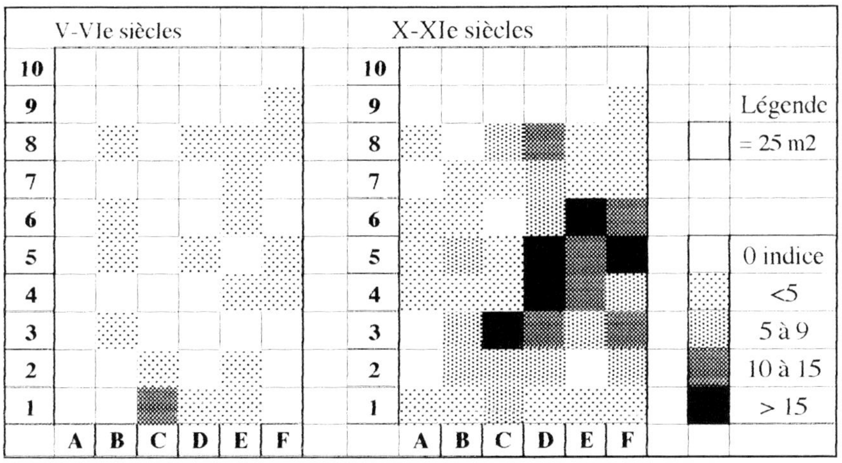

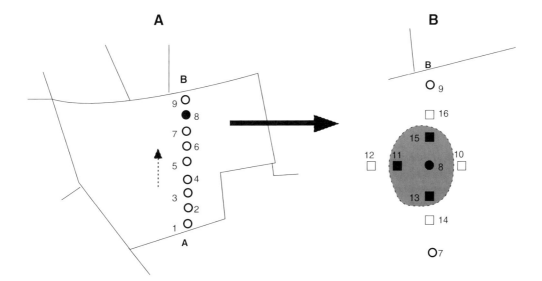

Figure 6.3   Languedoc. Prospection en carroyage sur l'établissement de Saint-Gilles-le-Vieux (Aimargues, Gard). Mise en évidence de deux phases d'occupation: occupation ponctuelle aux V–VIe siècles puis dilatation aux X–XIe siècles (mailles de 25 m²; doc. Cl. Raynaud).

Figure 6.3   Languedoc. Grid-prospection on the site of Saint-Gilles-le-Vieux (Aimargues, Gard). Demonstration of two phases of occupation: patchy in the 5th–6th century AD then expansion in the 10th–11th century (grid squares 25 sq m; source, author).

en carroyage s'impose particulièrement pour préciser la topographie des grands établissements à longue durée et la densité de leur occupation, qui connaît généralement d'importantes variations chronologiques, imperceptibles sans un enregistrement précis des éléments de surface (Figure 6.3).

Particulièrement adaptées aux terroirs de cultures annuelles ou de vignobles, ces techniques doivent être remplacées ou épaulées par d'autres approches dans les zones à herbages permanents, où l'on a pu mettre en pratique les carottages à la tarière, associés au dosage des phosphates à même de révéler — sous

Figure 6.4   Carottages à la tarière systématisés en milieu non labouré. Phase A: réalisation de carottages équidistants (20 m) sur l'axe de la bande de 300 m de l'autoroute A71 (point noir: carottage positif); Phase B: délimitation d'un indice de site à partir d'un carottage positif (point = carottage sur l'axe; carré = carottage complémentaire en vue de délimiter l'anomalie; point ou carré noir = carottage positif; Ferdière et Rialland 1994: 34).

Figure 6.4   Systematic coring with an auger in non-ploughed areas. Phase A: application of equidistant cores (20 m) on the axis of a 300 m strip of the A71 motorway (black circle = positive discovery); Phase B: site delimitation consequent on achieving a positive core discovery (circle = auger core on the axis; square = additional coring undertaken to find the boundaries to the anomaly; black circle or square = positive core result; Ferdière and Rialland 1994: 34).

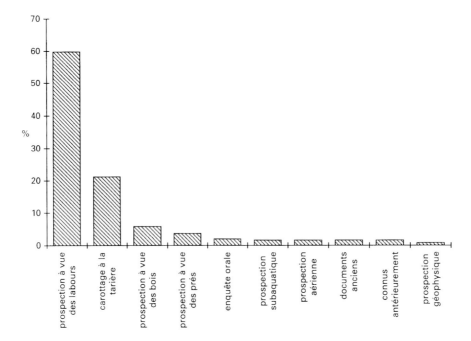

Figure 6.5   Berry, autoroute A71: pourcentage d'établissements révélés par les différents types de prospection. (D'après Ferdière et Rialland 1995: 52–53.)

Figure 6.5   Berry, A71 motorway: percentage of sites revealed by different types of prospection (after Ferdière and Rialland 1995: 52–53). From left to right: surface survey in ploughed land; subsurface augering; woodland survey; pastureland survey; oral enquiries; underwater prospection; aerial prospection; historic documents; known previously; geophysical prospection.

certaines conditions — la présence d'une occupation (Figure 6.4). Dans ces zones sans visibilité, on a aussi recours aux services de la photographie aérienne ou de la géophysique, dont l'efficacité n'est plus à démontrer en regard des révélations de sites mais qui demeurent inopérantes au plan de la datation. A l'évidence, ces méthodes se complètent plus qu'elles ne s'opposent, dans la mesure où, à l'exception des grandes plaines et des plateaux, les archéologues se trouvent généralement confrontés à une mosaïque de terroirs contrastés (Figure 6.5). La tendance consiste malheureusement à privilégier l'une ou l'autre technique, en raison des moyens disponibles et plus encore des spécialistes impliqués dans les équipes. Un travail d'intégration des techniques reste donc à promouvoir.

Une autre avancée sensible réside dans la mesure de mieux en mieux prise des singularités matérielles de chaque contexte culturel et/ou chronologique, qui font que l'on ne peut interpréter de même manière les chiffres concernant l'Age du Fer, la période romaine ou le Moyen Age (Zadora-Rio 1987: 14; Ferdière et Rialland 1995: 54; Raynaud; in press) (voir ce volume, Bintliff, pour la même observation).

En regard de la stratégie des projets en cours, la tendance majeure est à l'étude exhaustive de micro-régions, de l'ordre du canton ou de plusieurs cantons, de 100 à 1000 km². La pratique de l'échantillonnage

spatial (tirage au sort d'unités découpées arbitrairement) a été mise en œuvre occasionnellement, notamment en Bretagne, mais cette pratique demeure marginale (Langouet 1991: 40). Signalons toutefois une expérience originale par rapport aux modèles anglais en la matière. Il s'agit, dans la région de l'étang de Thau, sur le littoral méditerranéen, d'une grille d'échantillonnage systématique selon un maillage de 350 m. A chaque maille de la grille, un secteur-test de 100 m² fait l'objet d'une collecte exhaustive des artéfacts, qui permet ensuite d'analyser la densité d'occupation en s'affranchissant de la traditionnelle carte des sites et en raisonnant plutôt en termes de maîtrise spatiale et de territoires préférentiels (Figure 6.6; Bermond et Pellecuer 1997).

Représentativité des données de surface et problèmes de taphonomie

L'accroissement sensible du volume des opérations de prospection et de la finesse des enregistrements ont eu pour conséquence une confiance accrue dans une discipline longtemps décriée en France. Certains en viendraient même à penser — sans toujours le dire — que la prospection pourrait, dans certains cas, remplacer la fouille dans l'identification des établissements, et à moindre coût. On voit en effet se

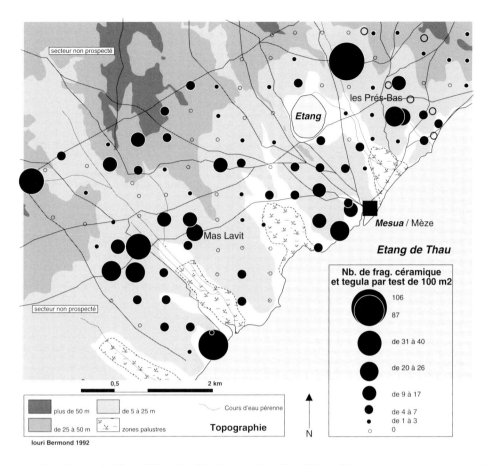

Figure 6.6   Languedoc, Bassin de Thau (Hérault). Résultats de la grille d'échantillonnage en carroyage, pour la période gallo-romaine (Bermond et Pellecuer 1997).

Figure 6.6   Languedoc, the Thau Basin (Hérault). Results from gridded sample squares (350 m on a side) for the Gallo-Roman period. Circle size reflects the number of pieces of pottery and tile for units of 100 sq m sampled per grid square (Bermond and Pellecuer 1997).

multiplier les exemples de prospection où la précision de l'enregistrement des matériaux de construction et des éléments mobiliers n'a plus guère à envier à celle d'une fouille (Langouet 1991: 44; Figure 6.7). Ces travaux, qui peuvent aller jusqu'à mettre en évidence le plan d'édifices, méritent encore que l'on s'attarde à leur évaluation et que des fouilles pratiquées à posteriori viennent contrôler leur degré de fiabilité. Je mentionnerai une expérience personnelle au cours de laquelle j'avais pratiqué un relevé de ce type qui dessinait le plan d'une petite ferme gallo-romaine, dans la région de Nîmes. La fouille est venue ensuite détruire l'hypothèse, révélant des aménagements en désaccord total avec l'image de surface (Figure 6.8). Je suis suffisamment, et depuis assez longtemps impliqué dans les travaux d'archéologie spatiale pour que l'on m'épargne l'accusation de porter discrédit aux techniques de prospection : je voudrais seulement insister sur le fait que l'archéologie de surface doit nourrir d'autres ambitions que celle de remplacer la fouille, combat perdu d'avance et dépourvu d'intérêt (Raynaud 1998).

Cette nécessité du *feedback* de la fouille constitue l'un des apports décisifs des grands travaux d'infrastructures, où prospection et fouille évoluent en étroite interaction. En pratiquant des fouilles extensives et en suivant systématiquement les travaux de terrassement, y compris hors des zones d'intervention archéologique, les équipes œuvrant sur les chantiers linéaires ont la possibilité d'observer régulièrement des vestiges profondément enfouis sous des dépôts sédimentaires récents, qui les rendent invisibles en surface (*Nouvelles* 1994). Ainsi dans le Berry, le chantier de l'autoroute A71 a-t-il révélé 14 établissements non observés par les prospections, soit 5.5% du total des sites répertoriés à l'issue de l'opération. Ces vestiges étaient masqués pour la moitié par des herbages, pour moitié par des dépôts alluviaux ou colluviaux sur des terrasses basses ou à proximité des cours d'eau (Ferdière et Rialland 1995: 65–66). Dans ce cas particulier, représentatif d'un climat tempéré et d'une topographie peu accidentée, la perte d'information pour la prospection demeure négligeable et les découvertes inattendues ne remettent pas

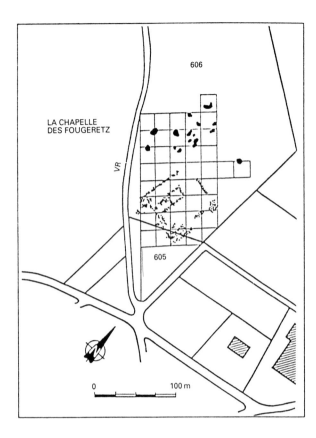

Figure 6.7   Bretagne, La Chapelle des Fougeretz. Mise en évidence d'un fanum par prospection en carroyage (Langouet dir. 1991: 44).

Figure 6.7   Brittany, La Chapelle des Fougeretz. Delimitation of a rural sanctuary (fanum) through gridded prospection (Langouet 1991: 44).

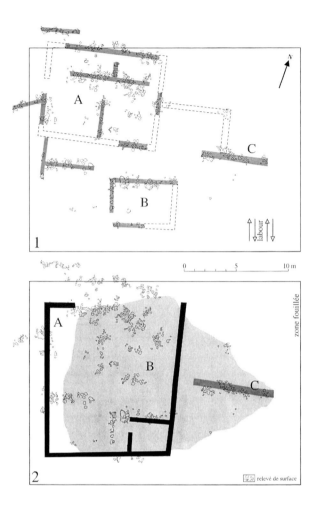

Figure 6.8   Languedoc, Saint-Côme-et-Maruejols (Gard). Relevé en carroyage des éléments de surface et restitution d'une ferme gallo-romaine (1); plan des vestiges après la fouille: un enclos de berger (2; doc. Cl. Raynaud).

Figure 6.8   Languedoc, Saint-Côme-et-Maruejols (Gard). Presentation of results from gridded surface prospection and reconstruction of a Gallo-Roman farmstead (1); plan of features after excavation: a pastoral enclosure (2; source, author.)

fondamentalement en cause l'interprétation des données de surface. Il en va bien autrement sous d'autres cieux... L'expérience récente des grands travaux linéaires en Provence et en Languedoc (TGV-Méditerranée, Gazoduc 'Artère du Midi') avec des sondages géo-archéologiques tempère cet optimisme en mettant l'accent sur l'épaisseur des recouvrements sédimentaires en piémont et dans les vallées, où les indices de surface ne peuvent prétendre traduire complètement les réalités antiques. Dans la moyenne vallée du Rhône, on a pu ainsi vérifier que 30–50% des établissements pouvaient se trouver enfouis sous 2–5 m de dépôts, ici principalement d'origine alluviale (Figure 6.9; Berger *et al.* 1997). Vient donc le temps des contrôles, par le sondage et par la fouille, par la mesure des processus d'érosion/accumulation. Cette approche des problèmes de taphonomie est encore trop rarement menée faute de spécialistes. Dans le cadre du TGV (train Lyon–Nîmes–Marseille), la phase des travaux préliminaires consistait à établir, avant la prospection, une carte d'évaluation de l'épaisseur des recouvrements susceptibles de

masquer les paléosols et les vestiges (recouvrement fort ou faible). La prospection de surface intervenait ensuite, sur les collines (Garrigue) et dans les basses plaines. Dans les vallées et en piémont, où les couvertures récentes peuvent atteindre 1 à 3 m, la prospection était ensuite réalisée à l'aide de tranchées, sur 5–10% de la surface concernée par le projet (Figure 6.10).

L'opération TGV-Méditerranée a fourni aussi l'occasion de développer la première application régionale des dosages géochimiques, sur un établissement romain à Lapalud (1995). Ultérieurement, la fouille complète du site a révélé une ferme gallo-romaine, confirmant l'analyse chimique. Cette

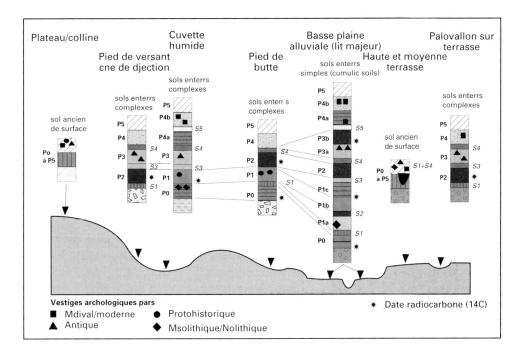

Figure 6.9 TGV-Méditerranée (train Lyon–Marseille). Etude géo-archéologique de l'épaisseur de la couverture masquant les paléosols et les artéfacts (Berger *et al.* 1997: 158).

Figure 6.9 The TGV-Méditerranée Project (section of railway improvement Lyon–Marseille). Geoarchaeological analysis of the thickness of cover masking palaeosols and artefacts (Berger *et al.* 1997: 158).

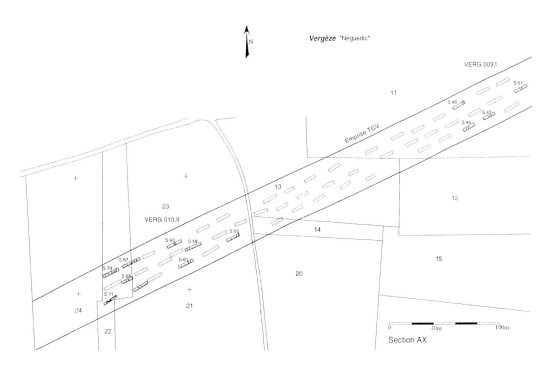

Figure 6.10 TGV- (train Lyon-Marseille). Grille d'échantillonnage en tranchées dans la vallée du Vistre, près de Nîmes (zone sondée = environ 10% de l'emprise). Bien que la prospection n'ait révélé aucune occupation, les tranchées ont mis au jour une ferme de la fin de l'époque romaine (en gris sur les tranchées, zone VERG 010, II), et un drainage agraire non daté pour l'instant (tranchées VERG 009, I; Roger 1996).

Figure 6.10 The TGV-Méditerranée Project (section of railway improvement Lyon–Marseille). Grid of trench sampling in the Vistre Valley, near Nîmes (zone tested approximately 10% of total). Although surface prospection did not reveal any occupation sites, the trenches brought to light a farm from the end of the Roman period (grey trenches, zone VERG 010, II), and agricultural drains so far undated (trenches VERG 009, 1; Roger 1996).

technique trop négligée a permis de pallier la dégradation des artéfacts par les labours modernes. Cette approche — peu coûteuse — mériterait d'être développée, spécialement sur les collines et dans les plaines médianes marquées par une érosion active, où structures et architecture sont généralement mal conservées.

Des typologies empiriques aux hiérarchies statistiques

S'il faut jalonner les voies futures de l'archéologie de surface — que l'on ne peut réduire à la seule prospection — c'est certainement dans la voie du calibrage de l'information et dans la quantification des données qu'il faut se diriger. Après avoir noté les progrès accomplis par les méthodes de terrain, effort qui doit être poursuivi, j'insisterai bien davantage sur les retards criants de l'interprétation des cartes archéologiques. Car enfin, j'observe avec inquiétude un divorce persistant entre d'une part des techniques et une stratégie d'acquisition des données sans cesse perfectionnées, et d'autre part des interprétations empiriques et routinières.

Les équipes méridionales ont tenté d'ouvrir une voie en ayant le souci de dépasser le stade de la carte archéologique pour tendre vers une intégration de l'information archéologique dans les champs croisés de l'histoire et de la géographie. En Languedoc, l'analyse statistique des données a été entreprise depuis le milieu des années 80, avec la double ambition d'une part d'enrichir le corpus des critères archéologiques en leur assurant une valeur statistique, et d'autre part d'insérer l'analyse des établissements dans une approche dynamique et spatiale, en caractérisant leur rôle respectif dans la structuration du paysage et leur position dans les réseaux de peuplement.

De nombreuses analyses multivariées ont donc été pratiquées, à la fois sur des échantillons régionaux différents et dans le cadre d'équipes diverses au sein du GDR 954 du CNRS, pour aboutir, en 1992–94, à une analyse portant sur près de 1000 établissements gallo-romains du Languedoc oriental et de la vallée du Rhône (Van der Leeuw 1995 ; Favory and Fiches 1994; Favory *et al.* 1995, Figure 6.11). Momentanément concentrée sur la période gallo-romaine, cette démarche conserve comme objectif d'appréhender progressivement les deux millénaires s'étendant de l'Age du Fer à la fin du Moyen Age. L'approche conjuguée des descripteurs de type archéologique et des descripteurs géographiques, croisés dans le cadre d'une Analyse Factorielle des Correspondances (AFC) et d'une Classification Ascendante Hiérarchique (CAH), méthodes complémentaires d'analyse des données (Sanders 1989;

Girardot 1983; 1995). Initialement calculées à partir de la carte, les données environnementales et sitologiques ont été ensuite élaborées à partir d'un Modèle Numérique de Terrain (MNT), puis intégrées dans un Système d'Information Géographique (GIS). Les *descripteurs archéologiques* rendent compte de l'information livrée par les gisements de vestiges affleurant le sol et par l'environnement: Superficie (6 modalités), Matériaux (6 modalités), Mobilier (4 modalités), Indices d'activité (7 modalités), Date d'implantation (14 modalités), Durée d'occupation (6 modalités), Occupation antérieure (3 modalités), Pérennité (2 modalités), Période d'occupation (13 modalités).

Ces descripteurs ont évolué vers un affinement des classes et une amélioration de leur représentativité. Ce processus a été encouragé par l'augmentation progressive du nombre d'établissements soumis à l'analyse, qui a autorisé la création de classes plus homogènes, tant au plan de la superficie des gisements, de la nature des matériaux utilisés dans l'architecture qu'à celui des découpages chronologiques. En revanche, en ce qui concerne la description du mobilier et l'exploitation des témoignages susceptibles d'éclairer les fonctions productives des établissements (Mobilier, Activité), on a dû, malgré des tentatives d'innovation, découragées par l'inertie des résultats statistiques, s'en tenir à l'organisation initiale des descripteurs. La pratique intensive de l'Analyse de Données a ainsi relégué au second plan ces descripteurs qualitatifs, peu discriminants en regard du classement des 'sites', tandis qu'elle soulignait l'efficacité d'autres critères, chronologiques en particulier.

Les *descripteurs spatiaux* recensent l'information au point topographique occupé par l'établissement, et ses connexions avec les réseaux de communication et de peuplement: Terroir (4 modalités), Sol (4 modalités), Pente (5 modalités), Distance à un cours d'eau (7 modalités), Distance à la voirie (6 modalités), Nombre de chemins (5 modalités), Nombre de relations avec les établissements contemporains (4 modalités). Ces descripteurs s'imposent à l'expérience parmi les plus discriminants.

Notre approche du rapport entre site archéologique et milieu environnant se fonde sur les analyses effectuées pour l'essentiel par UNISFERE (laboratoire d'analyse cartographique, Besançon). RAAP (Amsterdam) lui a fourni des fichiers élaborés par le SIG, à partir de données spatialisées et géoréférencées acquises sous forme de fichiers numériques, ou par digitalisation de données livrées par la cartographie disponible. En outre, ce laboratoire a calculé la distance de chaque site au cours d'eau ou au plan d'eau le plus proche et a mesuré le degré de la densité viaire, c'est-à-dire le nombre de chemins alentour du site, qui est un indicateur de la capacité des établissements à structurer l'espace environnant.

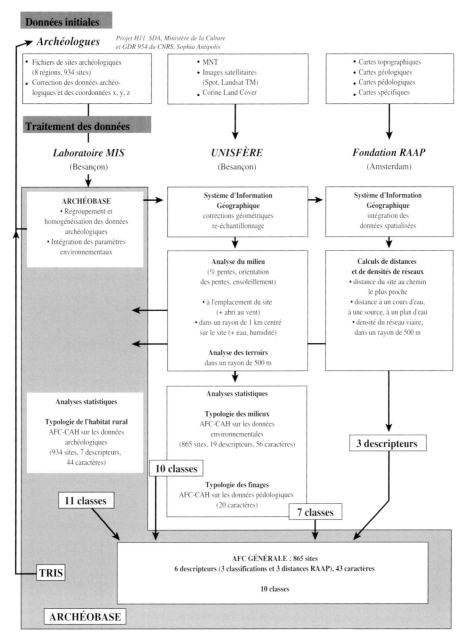

Figure 6.11 Vallée du Rhône. Diagramme méthodologique sur le croisement des données archéologiques et géographiques, dans le cadre du projet ArchaeoMedes (Favory *et al.* 1995).

Figure 6.11 The Rhône Valley. Diagram to illustrate the methodology for integrating the archaeological and geographical data within the ArchaeoMedes Project (Favory *et al.* 1995).

La description du milieu physique a porté sur deux de ses composantes, conçues à la fois comme les plus facilement exploitables à partir des sources documentaires disponibles et accessibles, et comme les plus stables — appréciation toute relative, bien entendu — ce qui importait dans une démarche qui prétend caractériser le milieu tel qu'il existait entre 21 et 15 siècles avant l'actuel. Il s'agit d'une part du relief et des effets qu'il induit, singulièrement du point de vue de l'exposition solaire et de l'exposition aux vents dominants, d'autre part du contexte pédologi-que, de stabilité variable selon les profils topographiques et hydrographiques considérés. L'analyse a été effectuée à deux échelles, selon les variables. Une partie d'entre elles porte sur l'emplacement même de l'établissement: pourcentage et orientation de la pente ou absence de pente, rayonnement solaire, exposition aux vents dominants. L'autre partie contribue à caractériser les alentours du site dans un rayon d'un kilomètre (proportions de secteurs plats et des différentes classes de pentes) ou de 500 m (associations de sols). Cette seconde classe de paramètres est chargée de décrire le cadre physique où s'exerce l'activité agro-pastorale et de suggérer, en respectant les compétences et les limites des forces productives de l'époque, une typologie qualitative des milieux exploités par les occupants des habitats étudiés. Les calculs ont permis de dégager une typologie des milieux topographiques (Van der Leeuw 1995: 143–68) et une typologie des finages occupés par les sites gallo-romains (Pédologie: 7 modalités).

Quelques résultats

L'analyse statistique multivariée appliquée aux 934 établissements de la base ArchæoMedes et aux seules variables archéologiques a confirmé la structure de la typologie issue de l'analyse consacrée antérieurement à 108 sites du Lunellois, en Languedoc oriental, ce que suggérait déjà la comparaison des tris à plat, donnant la fréquence de chaque variable. La CAH calculée sur les résultats de l'AFC répartit les sites en 11 classes, dont quelques sites fouillés éclairent la signification (Figure 6.12):

A (202 sites; 22%): très petits sites du Ier siècle, brève durée, vocation agraire sans habitat.

B (67 sites; 7%): établissements d'occupation supérieure au siècle, vocation agraire pour une partie, petits habitats pour une autre.

C (34 sites; 4%): petits établissements de facture indigène, à durée d'existence brève; habitats temporaires ou annexes agraires.

D (154 sites; 16%): petits sites de la fin de la République et du Ier siècle A.D., durée brève; aucun site fouillé n'éclaire cette classe.

E (125 sites; 13%): petits et moyens sites, de facture modeste, à durée excédant un à plusieurs siècles; fermes développées.

F (75 sites; 8%): petits sites tardo-antiques à occupation brève; annexes agraires.

G (85 sites; 9%): sites moyens tardo-antiques et de durée moyenne; petites fermes ou habitats temporaires.

H (75 sites; 8%): taille moyenne, implantés à la fin du Ier siècle B.C. et au Ier siècle A.D.; fermes indigènes aisées et petites villae.

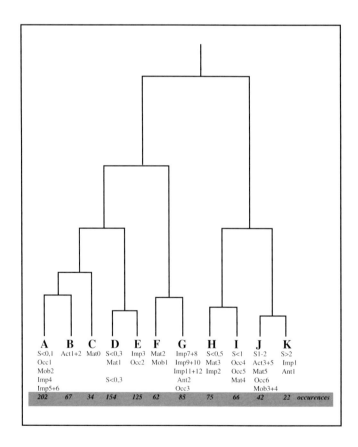

Figure 6.12 Vallée du Rhône, projet ArchaeoMedes. Graphe simplifié de la CAH (Classification Ascendante Hiérarchique), calculé sur les résultats de l'AFC (Analyse Factorielle des Correspondances), mettant en évidence la hiérarchie des établissements gallo-romains (Van der Leeuw 1995).

Figure 6.12 The Rhône Valley, ArchaeoMedes Project. Simplified graph of the CAH (Classification by Ascending Hierarchy), calculated on the basis of the AFC (Factor Analysis), revealing the hierarchy of Gallo-Roman sites (Van der Leeuw 1995).

I (66 sites; 7%): taille moyenne à grande, implantés à la fin du Ier siècle B.C. et au Ier siècle A.D.; villae.

J (42 sites; 4%): grands sites durables, confortables; petites agglomérations ou grandes villae.

K (22 sites; 2%): sites plus grands; grandes agglomérations ou grandes villae, dotées d'ateliers artisanaux.

Approche systémique du peuplement: des réseaux hiérarchisés

A l'issue de cette élaboration des données, l'archéologue dispose d'une information qui le met en mesure de raisonner à la façon du géographe, en termes de système d'organisation du peuplement. Au delà de la typologie des 'sites', l'analyse multivariée permet ainsi de préciser les modalités des processus de dispersion et de concentration des réseaux d'habitats, polarisés par des agglomérations ou par des établissements au statut architectural et fonctionnel supérieur (des villas?) à celui des établissements dispersés alentour. La projection sur la carte des résultats de l'analyse statistique met en lumière la cohérence du maillage hiérarchique.

Un *réseau d'habitat* se présente comme un dispositif spatial cohérent, regroupant à une période donnée des établissements aux fonctions diverses, liés entre eux par des relations fonctionnelles et hiérarchiques. La forme dominante de ces réseaux est structurée à partir d'un pôle d'initiative agraire, dont la fonction organisatrice est relayée par des habitats intermédiaires, hameaux ou fermes de taille moyenne et dont le contrôle et la maîtrise de l'espace agro-pastoral s'exercent concrètement grâce à des locaux et des aires techniques spécialisées, installées au plus près de l'activité de production, ainsi que par des petits habitats occupés de manière temporaire à l'occasion de certaines activités. On parle alors de réseaux polarisés. Initialement pratiquée de façon empirique par les seuls archéologues, la restitution de ces réseaux a fait l'objet d'une modélisation menée par une équipe de géographes de l'équipe PARIS (CNRS — Université Paris I) avec qui nous poursuivons une collaboration prolifique (Favory *et al.* 1998a; 1998b). Dans ce cadre, la convergence de la pratique empirique des archéologues et de la modélisation statistique des géographes renforce la valeur du modèle (Figures 6.13, 6.14).

Un second type de réseau, associant des établissements de rang sensiblement égal et n'ayant pas ou peu suscité l'émergence de 'sites' annexes, a pu être identifié sur le littoral Lunellois, autour de la Petite Camargue. En marge des réseaux polarisés et dans le contexte spécifique d'un littoral lagunaire, ce voisinage d'établissements de rang moyen évoque une

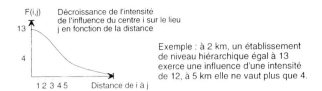

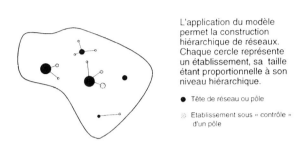

Figure 6.13 Languedoc oriental. Modélisation des réseaux locaux de peuplement (Favory, Mathian, Raynaud et Sanders 1998: 209).

Figure 6.13 Eastern Languedoc. Model of local population networks (Favory *et al.* 1998: 209). The fall-off graph with distance illustrates the decreasing intensity of influence from a centre i on a place j due to distance. Thus, for example, at 2 km a site with a hierarchical level (F) of 13 exercises an influence of intensity 12, but at 5 km it is worth no more than 4. The two-dimensional map shows how the application of the model permits the construction of a hierarchical network. Each circle represents a site, its width proportional to its place in the hierarchy. Dominant and subordinate sites are black and grey respectively.

économie domaniale, dans ce que l'on propose de qualifier, faute de hiérarchie explicite, de réseaux linéaires (Favory *et al.* 1994: 170–180).

Après cette phase d'expérimentation de la méthode, les équipes engagées dans le projet ArchæoMedes II (1997–99) se sont fixé pour objectif d'en tester l'efficacité sur l'Age du Fer et sur le Moyen Age.

Conclusion: vers l'intégration historique des données archéologiques

Avancée significative vers une analyse systémique du peuplement, cette perception des réseaux devra dans les études à venir rendre compte des différentes strates et stratégies d'appropriation du territoire, depuis le niveau de l'exploitation paysanne ou domaniale, jusqu'aux réseaux supérieurs exprimant la polarisation des campagnes autour de la ville. Les premiers essais de modélisation et d'analyse formelle de ces réseaux nous ont permis de mesurer toute l'efficacité d'une étroite collaboration avec les géographes.

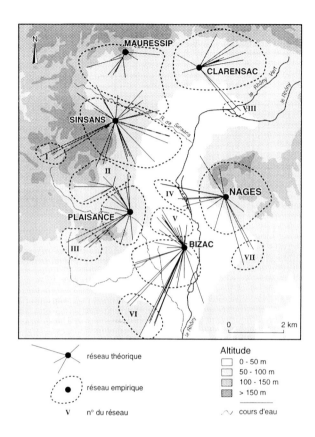

Figure 6.14 Languedoc oriental, région de Nîmes (Gard). Confrontation des réseaux empiriques (approche archéologique) et des réseaux théoriques (approche statistique). Globalement concordantes, les deux approches se distinguent par leur niveau d'insertion spatiale. La modélisation théorique opère sur les réseaux locaux, de l'ordre de plusieurs kilomètres d'extension, tandis que la modélisation archéologique aborde les réseaux micro-locaux, dans l'infra-kilométrique. C'est finalement la superposition des deux niveaux qui élucide la logique du système spatial, avec dans ce cas la soumission des unités domaniales (villas) aux places centrales que constituent les agglomérations (Favory, Mathian, Raynaud et Sanders 1998: 216).

Figure 6.14 Eastern Languedoc, Nîmes region (Gard). Comparison of empirical networks (archaeological approach) and theoretical networks (statistical approach). Overall in agreement, the two approaches can be distinguished by the level at which they are inserted into space. Theoretical modelling works on local networks of the order of several kilometres wide, whilst the archaeological modelling treats of microlocality networks at the subkilometre level. The logic of the spatial system emerges via the superposition of the two levels, with in this example the submission of the estate domains (villas) to central places that constitute the regional agglomerations (Favory *et al.* 1998: 216). Key: circle with radial lines = theoretical network; dashed bounded areas = empirical networks.

Si le déploiement des recherches semble bien engagé en de nombreuses régions, laissant augurer d'ambitieuses synthèses, il reste à promouvoir l'intégration verticale des résultats. L'approche systémique, telle que la pratiquent écologues et géographes, n'est pas encore une réalité en archéologie. L'organisation des équipes n'a pas encore pleinement intégré l'articulation des trois niveaux d'approche que constituent l'étude de site, la carte de peuplement, la restitution des réseaux. Force est de noter que trop de recherches sur l'occupation des sols demeurent démunies d'une analyse des réseaux parcellaires et de communication, et qu'à l'inverse certaines études de voies en restent à une approche 'monumentale', privilégiant un axe majeur en méconnaissant sa place réelle au sein de réseaux de peuplement et sa relation aux autres axes. De même les outils statistiques de l'analyse de données, appliqués ces dernières années à des études sur la vallée du Rhône (projet ArchæoMedes), restent à généraliser. Regrettons aussi que certaines études paléo-environnementales en habitat ne soient pas encore encadrées d'études territoriales à la hauteur de leurs ambitions. Un autre défaut de conception des projets d'étude est leur concentration sur une unique entité paysagère, au détriment des autres éléments du géo-système, négligé et méconnu: par exemple l'étude d'une vallée ou d'un delta, sans se préoccuper des reliefs en amont . . . .

La même carence concerne la faiblesse des études diachroniques, ciment nécessaire pour souder une communauté scientifique trop partagée entre protohistoriens d'un côté, antiquisants et médiévistes de l'autre. Milieux, peuplement et territoires ne s'étudient pas dans l'instant, les effets de tel ou tel phénomène — conquête agraire, déprise, mutation technique, pression démographique, etc. étant toujours différés. De ce point de vue, persiste un déséquilibre entre les études spatiales sur la période romaine et alto-médiévale, considérablement développées depuis une décennie, et l'impact encore limité de cette thématique en Protohistoire et au bas Moyen Age. Néanmoins, les discussions au sein des équipes CNRS/Culture qui maîtrisent les programmes révèlent un métissage à l'œuvre, la mesure des lacunes étant prise. Rien de bien nouveau au pays de Fernand Braudel…!

En définitive, plus que de nouvelles méthodes, l'archéologie spatiale en France a besoin de trouver un nouveau souffle, d'abord en harmonisant des pratiques encore trop disparates, ensuite en étoffant un cadre conceptuel très flou. Il est grand temps pour les archéologues de faire de la géographie!

References

Berger, J.-F., J.-L. Brochier, C. Jung and Th. Odiot
  1997 Données paléogéographiques et archéologiques dans le cadre de l'opération de sauvetage archéologique du TGV-Méditerranée. In *La dynamique des paysages*. 17e Rencontres Internationales d'Archéologie et d'Histoire d'Antibes, 155–83. Sophia: Antipolis.
Bermond, I. and Ch. Pellecuer
  1997 Villa et territoires en Narbonnaise: le domaine de la villa des Prés-Bas (Loupian) et l'agglomération de Mèze (Hérault). *Revue Archéologique de Narbonnaise* 30: 63–84.
Favory, F. and J.-L. Fiches
  1994 *Habitat et occupation des sols en France méditerranéenne, dans l'Antiquité et le Moyen Age. Approches régionales*. Documents d'Archéologie Française 42.
Favory, F., J.-L. Fiches and Cl. Raynaud.
  1998a La dynamique de l'habitat gallo-romain dans la basse vallée du Rhône. In Archaeomedes (collective publisher), *Des oppida aux métropoles*, 73–116. Paris.
Favory, F., J.-J. Girardot, Cl. Raynaud and K. Roger
  1994 L'habitat gallo-romain autour de l'étang de l'Or (Hérault). Hiérarchie, dynamique et réseaux du IIe s. av. au Ve s. ap. J.-C. In *Mélanges Pierre Lévêque* 8: 123–215. Paris: Les Belles Lettres.
Favory, F., J.-J. Girardot, Cl. Raynaud and F. Tourneux
  1995 Mobilité et résistance de l'habitat gallo-romain en vallée du Rhône: indicateurs de l'attraction ou de la répulsion exercée par le milieu? In *L'homme et la dégradation de l'environnement*, XVe Rencontres Internationales d'Archéologie et d'Histoire d'Antibes, 263–84. Juan-les-Pins.
Favory, F., H. Mathian, Cl. Raynaud and L. Sanders
  1998 Sélection géographique, déterminisme et hasard. In Archaeomedes (collective publisher), *Des oppida aux métropoles*, 151–248. Paris.

Ferdière, A. and Y. Rialland
  1994 La prospection archéologique systématique sur le tracé de l'autoroute A 71. 1ère partie. *Revue Archéologique du Centre de la France* 33: 7–86.
  1995 La prospection archéologique systématique sur le tracé de l'autoroute A 71. 1ère partie. *Revue Archéologique du Centre de la France* 34: 5–87.
Girardot, J.-J.
  1983 Micro-informatique et procédures conversationnelles des données. Le logiciel 'Anaconda'. *Cahiers de Géographie de Besançon* (Actes du 11e colloque sur les Méthodes Mathématiques Appliquées à la Géographie. Besançon) 25: 231–63.
  1995 Analyse statistique de l'habitat rural antique. Méthodologie. In Van der Leeuw 1995: 4–11.
Langouet, L.
  1991 *Terroirs, territoires et campagnes antiques. La prospection archéologique en Haute-Bretagne*. Revue Archéologique de l'Ouest, supplement no. 4.
Nouvelles,
  1994 L'archéologie préventive en milieu rural et ses phases d'évaluation, dossier collectif. *Les Nouvelles de l'Archéologie* 58: 1994.
Raynaud, Cl.
  1998 De la prospection à la fouille, et retour … Us et abus de la prospection méthodique: une expérience languedocienne. *Homo Faber* 2(1): 7–14.
  in press Définition ou hiérarchie des sites? Approches intégrées en Gaule Narbonnaise. In M. Pasquinucci and F. Trément (eds.), *Mediterranean Landscape Archaeology 4: Non-Destructive Techniques Applied to Landscape Archaeology*. Oxford: Oxbow.
Roger, K., A. Garnotel and G. Sachot
  1996 *TGV Ligne 5 — Secteur III: Avignon-Montpellier, Lot 42*. Rapport inédit. Montpellier: Service Régional de l'Archéologie.
Sanders, L.
  1989 *L'analyse statistique des données appliquée à la géographie*. Montpellier: GIP RECLUS.
Van der Leeuw, S. (ed.)
  1995 *Dégradation et impact humain dans la moyenne et basse vallée du Rhône dans l'antiquité*, I–II. (ArchaeoMedes Project 'Understanding the natural and anthropogenic causes of soil degradation and desertification in the Mediterranean basin', Vol. 3). Cambridge: Cambridge University.
Zadora-Rio, E.
  1987 Archéologie du peuplement : la genèse d'un territoire communal. *Archéologie Médiévale* 17: 7–65.

Outline: Environments, Population and Territories in Southern France. Spatial Archaeology at the Crossroads

Introduction

Since the 1980s, French archaeology has been progressively catching up with other countries in the study of settlement history and early territorial organization. The last decade has seen a redoubled effort to fill in the archaeological map, as much in the quantity of information as in the quality of the data registered (Figure 6.1). For long archaic, the methods for identifying and inventoring sites have progressed rapidly, borrowing from those of Anglo-Saxon projects (Figure 6.5) and yet adapting them to regional physical constraints — gridded collection (Figure 6.2: fieldwalking model; Figures 6.3 and 6.7: gridded collection), a rigorous calibration of artefacts (Figure 6.6: ceramic and tile density), test coring in a grid (Figure 6.4) and the confrontation of surface finds with excavation (Figure 6.8). This growth has led to a strong support for this young discipline (too rapidly?), giving to its practitioners a great confidence (too great?) in their results.

Collection and interpretation of spatial data

Our recent experience with major linear construction programmes in Provence and Languedoc (the TGV, the 'Midi Artery' Pipeline) and geoarchaeological investigations, has blunted this optimism by stressing the thickness of sedimentary cover in the piedmont and valley-bottom zones (Figure 6.9), where surface traces cannot pretend to reflect ancient realities. We are now required to find controls, by test-pits and excavation, through measuring the processes of erosion and accumulation (still too rarely carried out) (Figure 6.10). And to avoid remaining literally 'superficial' our surface archaeology must be linked more regularly to geophysical prospection — underdeveloped due to a shortage of such specialists across the landscape.

From empirical typology to statistical hierarchies

Statistical analysis of French settlement data has been undertaken since the middle of the 1980s, with the ambition, on the one hand, of enriching the corpus of archaeological criteria through ensuring their statistical validity, and, on the other hand, to inject the analysis of sites into a dynamic spatial approach, through characterizing their role in the structure of the countryside and their position in networks of population.

Numerous multivariate analyses have recently been undertaken using different regional samples and from diverse projects, leading by 1994 to an analysis covering almost 1000 Gallo-Roman sites in E. Languedoc and the Rhone Valley. Although at the moment concentrated on the Gallo-Roman period, this approach has as its objective to incorporate progressively the 2000 years from the Iron Age to the end of the Middle Ages (Figure 6.11).

The approach marries archaeological type descriptors and geographical descriptors, linked within the framework of a correspondence analysis and a classification of ascending hierarchy — complementary methods for data analysis. Initially calculated from maps, the environmental and site data have later been developed within a digital terrain model before integration into a GIS. *The archaeological descriptors* take account of the information provided by the site deposits and its geography: surface area, building materials, artefacts, activity traces, date of foundation, length of occupation, previous occupation, seasonality, and period of occupation. These descriptors have evolved towards a refinement of classes and improvement in their representativity. This process was encouraged by the progressive rise in the number of sites submitted to the analysis, allowing us to create more homogeneous classes using not only chronology but also site plan and construction materials. Looking back, as far as concerns the description of artefacts and the study of evidence for the productive function of sites, the inertia of the data meant that we kept to our initial descriptors. This means that qualitative descriptors have taken second place as useful discriminators to other criteria, especially chronology. *The spatial descriptors* summarize information regarding the position occupied by the site and its connections with population and communication networks: territory, soil, slope, distance to routes, number of routes, and number of links with contemporary establishments. These descriptors, in our experience, were amongst the most useful discriminants. In calculating these factors, digital and GIS technology played a major role. The description of the physical surroundings has rested on two components — seen as the most easily accessible from our available documentation and as the most stable — (a relative set of values, however!) and this was focused on a characterization of the environment as it existed between 21 and 15 centuries before the present. On the one hand, it looked at the effects of relief, especially insolation and relation to dominant winds; on the other, it looked at pedology (where stability varied with topographic and hydrological factors). This analysis was set at two scales: the first focused on the immediate site location: percentage and orientation of the slope or its absence, solar orientation, wind exposure. The second characterized the site surroundings within a 1 km radius (proportion of flat sectors and of different slope classes) or to 500 m (soil associations). This latter group of parameters describes the physical framework for agro-pastoral activities and, allowing for the constraints of past productive technology, can help create a qualitative typology of the environments exploited by the occupants of these sites. Calculations have led to a typology of topographic associations and a typology of the territories occupied by Gallo-Roman sites.

Some results

Multivariate statistical analyses of 934 sites within the *ArchaeoMedes* Project, looking just at the archaeological variables, has confirmed the typology of sites that resulted from the earlier analysis of 108 sites in E. Languedoc, encouraging a large-scale grouping exercise. Eleven classes of Gallo-Roman sites emerge, with excavated sites helping to clarify each class (Figure 6.12).

Systemic approach: hierarchical networks

Beyond the typology of sites, multivariate analysis allowed us to focus on the patterns of dispersal and nucleation of settlement systems, especially the polarization between agglomerations or monumental sites of superior role (villae?) and lesser sites dispersed around them. Projecting back on to the map the results of the statistical analysis emphasizes the coherence of the hierarchical network.

A settlement network appears when we see a coherent spatial cluster where, for a particular period, sites of diverse function can be linked by functional and hierarchical relationship (Figures 6.13 and 6.14). The dominant form of these networks appears as a focus of agricultural initiatives, whose organizing role is relayed through intermediate settlements —hamlets or medium-sized farms, and whose mastery of the agro-pastoral space is exercised concretely via localities of specialist activity — or by small settlements occupied temporarily for certain activities. We term these settlements Polarized Networks. A second type of network, associating sites of equal rank and lacking annexe-sites, has been identified in the Lunellois region. On the margin of the polarized networks and in the specific context of a littoral lagoon, this district of middle range sites evokes an estate economy — and lacking explicit hierarchy we call these Linear Networks.

Historical integration: the progress of Gallo-Roman colonization in Southern France

The phase of the creation and diffusion of both dispersed and grouped settlements reaches its peak in the 1st century AD, following rapid growth, then the number of sites tends to be reduced from the 2nd century AD. Population decline is checked, in certain regions, by a second phase of site creation in the 4th–5th century AD.

If we project the regional statistics on to the map we see that the most precocious development of Romanization operates in the lower Rhone Valley, doubtless under the influence of the Greek colony of Marseilles, allied with Rome. Next, the agrarian colonization front progresses along the Rhone Valley up to the end of the 1st century AD. It is those regions that see the latest rise — marginal zones in relation to the geographic core of Romanization, where we shall see a revival of site creation in the Late Empire.

What the general curves show, among other things, is that site *size* is only a partial indicator of functional hierarchy: site status has to be tied to a specific occupation history. Thus small sites do not evolve like medium-sized or large sites. The archaeological criterion of size then only has value if related to the date and occupation length of a site. In progress are analyses of different strategies of land use, from peasant production to the estate mode, and modelling the networks around towns.

Although geographical approaches are seen here to be widely used in numerous regions, pointing the way to ambitious syntheses, we still need to promote the vertical integration of the results. A systematic approach imitating those of ecologists and geographers is not yet a reality in archaeology. Projects have not fully integrated the three levels of approach that comprise the study of the site, the map of population distributions and settlement networks. Too many analyses of settlement in the landscape lack the study of land allotment or communication networks — and, on the other hand, certain studies of routes have taken a 'monumental' approach privileging a major axis and neglecting its real place in the settlement system. I also regret that certain palaeoenvironmental studies have excluded territorial analysis from their objectives. Another failing is the overconcentration on littoral areas to the detriment of a neglected and poorly known hinterland.

A similar deficiency meets us with the poverty of diachronic studies, an essential cement for a scientific community that is too segregated into prehistorians and students of Antiquity. Milieux, population and territory are not concepts to be studied in one moment of time — nor their effects a single phenomenon: agrarian conquest, or depression, technological change, population pressure — these fundamental variables rarely have the same expression. From this point of view there is also a disequilibrium between spatial studies in the Roman and Early Medieval eras — considerably advanced for a decade now — and the limited impact they have had in Protohistory and the High Middle Ages. However new projects promise to improve on these weaknesses.

In conclusion, rather than new methods, spatial archaeology in Southern France must get its breath back, first, through harmonizing practices that are still variable, second, in extinguishing some rather hazy concepts and attitudes to time. The existence of a scientific community that is held together and regularly revitalized by young researchers augurs well for dramatic changes in the near future!

Editor's Note: The English version of this paper was prepared by John Bintliff from an earlier draft of the chapter and both includes additional material and excludes some of the material in the French version.

7. The Past, Present and Future of the *Polish Archaeological Record* Project

Paul Barford, Wojciech Brzeziński and Zbigniew Kobyliński

Summary

This article discusses a unique nationwide programme, run by the State Service for the Protection of Monuments in Poland, for the systematic inventorization of archaeological sites based on extensive archival research and systematic fieldwork. The history of the development of the programme and its basic principles are presented. The methodology of conducting the work is outlined in some detail, as are some of the technical and methodological problems that have arisen in the course of the compilation of the record. The use of the records for research is discussed, as also is their use in the process of heritage management and public education. The relationship of the record to other types of survey technique is outlined as is its computerization and dissemination. The future of the project is also briefly discussed. The Polish Archaeological Record is an advance on many other European SMRs, since it is based on active systematic and detailed data collection that will eventually cover the whole country.

Aims of the project

This paper describes and discusses a programme of producing a homogeneous national archaeological Sites and Monuments Record (SMR) for Poland that has been operating since 1978. This Polish Archaeological Record (in Polish 'Archeologiczne Zdjęcie Polski', which is commonly abbreviated as AZP), an ambitious and idealistic undertaking to accomplish a national database of archaeological sites, is state financed (by the Ministry of Culture and Arts[1] through the Office of the Conservator-General) and is also supported by provincial governments and communal authorities. The record is compiled by systematic fieldwalking of the entire country within a grid of arbitrarily defined areas and by systematizing the archival information, museums collections and bibliographic references. The systematic and detailed field prospection is carried out according to standardized procedure and with the use of standardized recording sheets.

The AZP has three main aims:

1) scientific recording (accumulating evidence of the real numbers and locations of sites in the landscape);
2) conservation (collecting of information about the location of sites to allow planning of activities taken to combat actual and future threats);
3) education (adding to public knowledge and appreciation of the early history of Poland).

This project, unique in its scope and execution, seems to be in advance of other European national and regional sites and monuments records. From this aspect it is perhaps a fitting subject in a volume dedicated to the future of fieldwalking. This account supplements those published elsewhere in western languages (Jaskanis 1987; 1992; Konopka 1984b).

Estimates of the time required to survey the whole country (312,520 sq km) suggested it would take about 15–20 years to complete (Kempisty *et al.* 1981: 23). Now exactly 20 years after the beginning of the project, we are still engaged in completing the last 30% of the survey (Figure 7.1). This anniversary seems to be an appropriate point to describe this project, its achievements and failures, as well as presenting suggestions for the future.

While the project itself (still in the process of the completion of the first stage of the fieldwork) cannot be changed in mid-operation without affecting the comparability of its results across the country, one may now start to consider ways in which the project — in the light of experience — could in future be usefully modified to become a more effective tool in other areas. The process of assessment is aided by the recent publication of an attempt to summarize the progress of the project (Jaskanis ed. 1996).

History of the project

The history of the AZP project is summarized in Polish in several articles in Konopka 1981c (especially Konopka 1981a: 28–37; see also Jaskanis 1992: 81–82). It did not arise in a vacuum, Polish archaeology having over the past half-century paid great attention not just to individual sites, but also to their place in the overall settlement networks, their location within the natural environment and relationship to each other. In this respect, settlement studies in Polish archaeology developed under the influence of the works of German archaeologists, particularly Herbert Jankuhn and the group gathered around the journal *Archaeologia Geographica*. This approach put particular stress on constructing settlement maps and relating them (without, however, any discussion of biases inherent in such maps) to the reconstruction of the natural environment in prehistory.

As in many other countries in the inter-war period, several archaeologists in Poland — most notably Professor Józef Kostrzewski of Poznań — employed on a large scale techniques of looking for and recording new sites by searching ploughed fields and other subsoil exposures, and localizing and collecting artefactual material. Most of these excursions led to the discovery of sites in the more predictable locations, along river valleys, often for pragmatic reasons clustering near to railway stations, and reflecting the areas of activity of individual researchers, university chairs or museums. The distribution and type of finds reported to museums and other institutions were uneven and depended on a variety of associated circumstances, as did the recognition and degree of recording of earthwork sites. All of these factors introduced a bias into the picture of ancient settlement of the areas that now form the modern state of Poland, as seen from the viewpoint of the archives of the conservation services.

Another problem was the territorial changes in the period 1918–45 and the destruction or removal of museum resources and archives at the end of the Second World War, which created additional difficulties in the compilation of the basic information of where sites had previously been found and excavated. This information, together with new discoveries, was held in the SMRs in the archives of the conservation services and formed the basis of our knowledge of the location and distribution of archaeological sites across the country.

These problems are not restricted to Poland of course, but in the 1970s several voices were raised in Poland, pointing out the unsatisfactory information base from which the existing SMRs had been compiled, and leading to calls to produce information of a totally different order. This was therefore the background to the rise in the latter part of the 1970s of a project that was to have a fundamental effect on our knowledge of the distribution of ancient settlement in the area of modern Poland.

The fundamental principles of the AZP project seem to date back, however, to a tersely formulated article of one of the father figures of Polish archaeology, Józef Kostrzewski (1950: 19). He stated the need to involve many people in a long-term project that would collect information from archives

and museum collections and from landscape features (strongholds, barrows, cemeteries), but also surface finds and oral information as a result of fieldtrips. This work was to be on the scale of small regions until the mid-1970s. However, in the late 1960s and 1970s several large-scale fieldwork projects combined with museum work were compiling detailed settlement maps for several macroregions (e.g. T. Wiślański's work on the Neolithic of northwest Poland (1969), J. Kruk on the Neolithic of the loesses of southeast Poland (1973) and Z. Hilczerówna (1967) on the Early Medieval settlement in west-central Poland). These studies were concentrated on problems of settlement in particular periods in specific areas, but they revealed the increase in our knowledge of the picture of the distribution of settlement and number of sites that could be attained by detailed study of large areas. As an example, the detailed fieldwalking of the Pruszków area just to the west of the capital of Poland by S. Woyda (between 1966 and 1975) led to the discovery of considerably more sites than previous records had suggested (the survey increased the number of sites known from 80 to over 1000, including the discovery of a massive previously unknown complex of hundreds of iron-smelting sites).

An early project of the AZP was compiled in 1975 by investigators from several institutions (the so-called 'Project 1975'; Kempisty *et al.* 1981), but its realization had to wait another few years of detailed preparation. This project contained the main elements of the AZP project. A 'rival' project was compiled by the Poznań branch of the State Conservation Workshops (PP PKZ) in 1977 on a basis of the fieldwork that was being conducted by them in advance of major industrial developments. As a result of lengthy consultations, a common standpoint was reached on many issues, the Ministry of Culture undertook to finance the venture, and in April 1978 the enormous AZP project was born with the object of cataloguing all the archaeological sites in Poland (Konopka 1984b).

Overall control of the project was exercised by the Centre for the Documentation of Monuments (ODZ), which functioned in accordance with a decree of the Minister of Culture and Arts (30 December 1987; with changes introduced 22 December 1990) and on the basis of this (Section 2) has the statutory function of creating a central sites and monuments archive and of overseeing its correct utilization. The archaeology section of ODZ was set up — since 1981 directed by D. Jaskanis — primarily to service the project. A central archive containing the results of the AZP survey (supplementing that in regional centres) was created in the ODZ office in Warsaw.

From the inception of this project, many — mostly younger — archaeologists have done systematic fieldwalking all over Poland every spring and autumn, recording sites and the artefacts found on the surface

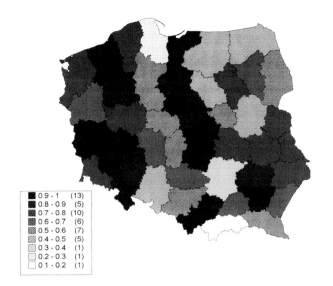

Figure 7.1   Completeness of the AZP project in various provinces of Poland (situation for 31 December 1997, according to data from the Centre for the Documentation of Monuments [ODZ]).

taken to ensure its continuation. In 1995 the post of Chief Archaeologist within the Office of the Conservator-General was created, and one of the first things he did was to force the recognition of a special status for the AZP project as a ministerial project of special significance to the construction of national conservation and research policies. Since the end of 1995 the programme has become a special programme of the Ministry of Culture and Arts, administered by the Office of the Conservator-General.

The attempt to investigate as many AZP areas as possible in the next few years is now seen as an extremely urgent priority. In the year 2000 all local government development and land use plans in Poland are due for revision, and the incorporation of the data from the SMRs into them is one of the most effective ways of controlling by administrative means the rate and process of destruction. The creation of a special government programme has thus had the result of a seven-fold increase in the financing of the project compared to the level in 1995. Together with the decrease at the beginning of the 1990s of the central funds for the AZP, the significance of local sources of financing grew (Figure 7.3). In 1993, for example, the funds of the Ministry of Culture and Arts accounted for some 70% of the AZP budget; in 1994 they amounted to just 53%; and in 1995 they fell to 47%. This lack was supplemented primarily by funds from the provincial councils and also funds from other institutions, such as museums, local governments and developers.

In 1996, however, central government funds again formed 90% of the financing of the AZP; this was

according to a predetermined methodology. In the years 1978–96 over 5535 of the 5 × 7.5 km search modules, into which the country has been divided for the purposes of the survey, were completed (an average of 369 per year). In the latter years of the 1980s and in the 1990s, with increasing economic problems of the communist state, the project was less well-financed by the central conservation service and was slowly grinding to a halt (Figure 7.2). The crisis in the AZP programme required new measures to be

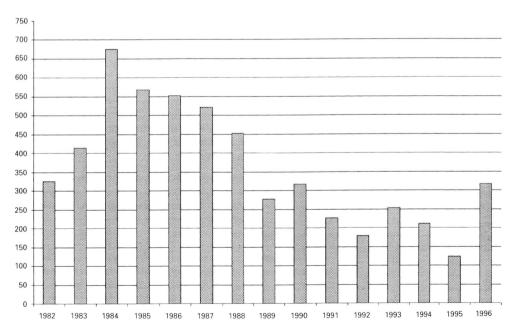

Figure 7.2   Number of the AZP modules searched in 1982–96: vertical axis (according to Jaskanis 1996 and the data provided by the ODZ). Fieldwork season: horizontal axis.

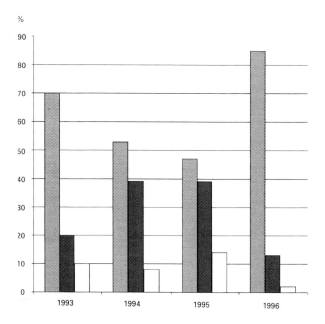

Figure 7.3   Sources of financing of the AZP project
(according to the data provided by the ODZ).
■ Service for Protecting Historical State
Monuments. ■ Provincial governments.
□ Other.

due to the formation of the government programme
administered by the Office of the Conservator-
General. Despite the strong financial involvement of
a number of provincial councils in the realization
of this project, it was not possible to avoid a drastic
drop in the number of examined search modules in the
period considered. Thus in 1995 the documentation
from only 124 modules was sent to ODZ. It is enough
to say that, in the middle of the 1980s, about 500–600
modules were being completed yearly (Figure 7.2).
A clear improvement was visible only in 1996 when
the AZP was reactivated as a government project
administered by the Office of the Conservator-General. In that year documentation from 317 search
modules was sent to ODZ. If it proves possible to
maintain the financing at its present level, it should be
possible to complete the first phase of the AZP in
the next eight years.

There is considerable variation in the density of
sites across Poland. Densities may vary as much as
from an average of 13 sites per module (Bielsko-Biala
Province) to 115 (Poznan Province), with an average
52 sites per module for the whole country (Jaskanis
1996: 35, tab. 11). On fertile soils there are individual
search modules with over 300 sites (for example in
Przemysl, Cracow, Tarnobrzeg and Kielce Provinces).

Theoretical assumptions

Despite the number of archaeologists who every year
systematically fieldwalk all over Poland and record

sites and artefacts found on the surface, there is little
concern with theoretical considerations among Polish
archaeologists involved in the project. The problems
of archaeological visibility of sites on the surface,
the techniques of their discovery and recording and
the optimal sampling strategies in settlement studies
have been discussed by some field archaeologists
(e.g. Kruk 1970; Mazurowski 1980; Kobyliński 1984;
Brzeziński *et al.* 1985). The results of the AZP project
are unfortunately quite often misused and unreasonably identified with the real structure of past settlement of the area, without taking into account the
conditions of doing the fieldwork, the pattern of
modern land use and erosion revealing sites, or hiding
them. Some of the theoretical assumptions of the
project seem to be worth discussion here.

The concept of the archaeological site

The basic task of the AZP according to its initiators
was the recognition and analysis of the spatial aspects
of the reserves of archaeological data in Poland.
A key question was therefore determining what was to
be the primary identifiable spatial unit of this
database. The central concept of the AZP project is
that of the archaeological site, which, although
apparently nowhere formally defined as such in the
Polish archaeological literature, is usually conceptualized as 'a spatially definable area where archaeological data [read: 'artefacts'] are found in association
and surrounded by an area from which such data are
absent' (Konopka 1984a: 28–29). The basic form of
documentation thus becomes the Archaeological Site
Record Card (*Karta Ewidencji Stanowiska Archeologicznego* [KESA]). Although such an approach seems
to be understandable and — perhaps — the only
possible at the time of the designing of a database
of the archaeological resources of a given country, it
raises certain doubts from the point of view of
archaeological theory. Observation of human behaviour, particularly in such densely settled areas as
we have in most parts of temperate Europe, shows
that the space utilized by man does not comprise only
two categories of areas — utilized and non-utilized.
In these regions, alongside those fragments of space
which are intensively utilised for settlement — farmsteads, villages, towns, one can demonstrate areas
that were utilized economically or were penetrated
sporadically for various reasons. Each of these areas
can contain archaeological data — material traces of
the presence or activity of man, though we know
from excavations that the number of finds in a given
place does not always correspond in a direct fashion
to the intensity or longevity of the use in the past of
the place by man. Archaeological sites can represent the place of permanent habitation, temporary
or cyclical habitation, or the place of a one-off

campsite, a place of burial of the dead, a place within an area utilized agriculturally (a broken pot that had been used to take food into the field) or for hunting (point of a lost arrow), a point on the route of human wanderings (objects dropped or lost on the way), and so on. In the Anglophone literature therefore there has been propagated a different approach to the relationship between man and space, known as 'nonsite' or 'offsite archaeology' (Bintliff and Snodgrass 1988). In the AZP project each individual find becomes a 'site'. Thus, for example, Matoga (1996: 49) recently suggests understanding the concept of site as a purely conventional category (investigative site) and not representing any fragment of past sociocultural reality.

Relation between the site and surface material

The basic assumption of the AZP project is the existence of a positive relationship between archaeological finds found on the surface of the earth and that which is to be found under it. In conceptual terms at least the assumption that the scatter of finds on the surface represents an archaeological site seems obvious. These are usually the scattered remains of archaeological deposits that have been shattered by the destructive action of ploughing (or the activities of burrowing animals). The presence of archaeological artefacts on the surface in a particular findspot could actually, however, result from a number of factors. Above all the processes of their deposition could result from deliberate discarding in the fields beyond the area of habitation and the area intensively exploited (for example, the practice often noted of Polish peasants throwing broken potsherds on to roads; Kobyliński and Kobylińska 1981: 52), accidentally lost during journeys, and so on. The surface archaeological material is also subject to various post-depositional and cultural disturbing processes. Archaeological artefacts can be carted on to the fields together with natural fertilizing materials, brought together with earth from another place in order to level the ground or to enrichen infertile soils (one of the authors recently had this experience in fieldwork near Landshut in Bavaria, where he was informed by the farmer of the true origin of sherds in dark soil on a field surface) and need not represent archaeological sites at all. Intensive ploughing may move artefacts in relation to their original position. They can be depleted by collectors or moved by playing children.

Number of artefacts on the surface and the definition of a site

These problems were realized at the beginning of the AZP project. In order to resolve them, a differentiated terminology was introduced for archaeological finds of differing quantities of artefacts. The term which was introduced was the so-called 'settlement trace' (*ślad osadnictwa*), a term introduced for 'a specific situation when the quantity of data is low, or occurs in such a low density that the definition of a site is impossible' (Konopka 1984a: 29).

The acceptance of the criterion of quantity of finds in the differentiation of settlement sites from 'trace sites' is, however, clearly fallible. From practical experience one may state that there is often an inverse relationship between the state of preservation of the below-ground part of a site and the number of finds on its surface. The less damaged the site is by ploughing or other human activity, the less fragments of ceramic or other artefacts appear on the surface. Good examples of this can be seen in the case of the settlements near the Early Mediaeval strongholds at Wyszogród, Płock Province or Haćki, Białystok Province (excavations by Z. Kobylinski), where, despite many years' excavations of the strongholds, nothing was known of the existence of the adjacent well-preserved open settlements until deeper ploughing revealed a few fragments of ceramics. A similar situation existed during the rescue excavations preceding the construction of a gas pipeline in Białystok Province in northeastern Poland (excavations by D. Krasnodębski). Despite repeated surface prospection, some sites were revealed as surface scatters of only a few finds, while investigations showed that they were well-preserved cemeteries or settlements. We may conclude, therefore, that a 'trace site' may be the indicator of a well-preserved site. However, it may equally be an indicator of a single event of little historical significance (for example, the breaking of a pot by a cow in a field, as Professor Waldemar Chmielewski of Warsaw University used to joke) and having no reflection in the form of an archaeological site under the ploughsoil.

Visibility

The visibility of surface archaeological material is in principle good over about half of the country (about 50% of the country is under the plough), however, across this area the visibility of sites is also variable, depending on the character of the crop and the season of year when the prospection is carried out. It was thus optimistically assumed that the search teams would carry out the fieldwork several times, which in practice did not take place. In other parts of the territory, and especially in forests (28% of the surface area of Poland), the observability of sites is very restricted, with the exception of barrows and other sites with a visible surface relief. Forests are very difficult to search, and often the effort is not rewarded by the discovery of sites, and by the same token does not earn money for the search team. In some provinces, therefore, forests are ignored altogether and their area is deducted from the area investigated in calculating payment, or the payment is halved for forested areas on the assumption that they are penetrated in a summary fashion. Generally, however, searching forested areas is less attractive, and provinces with large areas of forest have problems finding archaeologists willing to conduct these surveys. It is a similar case with meadows, which should be carefully checked, although it is known that that the chances of finding artefactual material on their surface is slight. The only chance of finding artefacts is by the searching of molehills, though the definition of the boundaries of a site on this basis is virtually impossible. Extremely difficult for prospection are urbanized and industrialized areas around (and within) towns, where every small garden (often surrounded with a fence) has to be found and investigated. In such a region a map of surface finds is usually rather a map of areas available for observation than an actual map of historic settlement (Horbacz and Olędzki 1994). Of course, sites covered by hillwash etc. are also impossible to discover, though they are well-protected and only threatened by developments involving earthmoving.

Thus negative results of prospection need not always mean the lack of a site in a given location (Lloyd and Barker 1981). In the literature may be found references to areas that have been repeatedly searched over periods of several years or longer periods. Czerniak (1996: 42), for example, investigated an area of 12 sq km six years after his first fieldwalking. The first season revealed 56 sites, the second 61, but including 34 new sites. Unfortunately, Czerniak does not describe this experiment in enough detail to assess its validity, nor the reasons why this effect occurred in this particular situation. Bienia and Żółkowski (1996) found that the database was enlarged by new sites (24%) in the area they verified, but that 16% of previously found sites were not located a second time. Banasiewicz (1996: 93) checked a previously surveyed area after 13 years and was unable to relocate 10% of the earlier sites, but found 30% new ones. The causes of such phenomena may be various. They may reflect a factual state of affairs — some of the sites previously visible may have been destroyed (as Banasiewicz intimates), and new ones revealed by plough-erosion. Some of the differences may be due to subjective or temporary factors, such as changing availability of parts of the area for observation, altered agricultural regimes, or even the method of conducting the fieldwork.

These difficulties in considering a single pass of the AZP search team as representing the real distribution of sites in an area result in the constitution of the view (Kurnatowska and Kurnatowski 1996: 81) that AZP is not a tool for the discovery of all sites, but rather for the discovery of zones of settlement.

Differential density of artefacts within a site and the problem of its boundary

We cannot know what we are in fact defining when we define archaeological sites on the basis of archaeological material on the surface, that is, what type of past phenomena we are identifying on this basis. A good example is the distribution of Late Mediaeval and Post-Mediaeval sherds found scattered in the fields around villages. When we define the boundaries of these scatters and according them the status of archaeological sites, we are actually not identifying the area of the former village itself, but demonstrating the extent of the carting of manure to the fields, or the extent of the throw of broken pottery of the owner of a farmyard or his children.

The problem of the definition of the boundaries of a site described above is still unresolved today. It is reflected in the number of sites identified by different investigators on the basis of surface material. A frequent case is the differential density of material on the surface. As one passes across a field from the centre of a site, one observes a decreasing density of finds until their disappearance, but one can observe an increase in their number again a few steps further on. Such phenomena often occur on fertile soils along watercourses (for example, on the Upper Vistula near Cracow, or on the Isar near Landshut in Bavaria). Do we define one site of elongated form, or rather several sites defined on the basis of the denser concentrations of finds? Two scatters of finds may be divided by a modern road, or by an area of unploughed land, making direct inspection of their continuity (or discontinuity) impossible. In such a case, it is a question of individual choice whether we will state the presence of several sites, or only one. Another difficult case is the very frequent distribution of Late Mediaeval and Post-Mediaeval ceramics and other finds around modern villages of earlier foundation. The differentiated possibilities of conducting observations of these scatters often mean that the material is visible only in a few fields, in between which are orchards or farm buildings. These scatters probably form one extensive spread, but one cannot prove this, and thus many investigators define such a site as a collection of smaller ones on the basis of their observations.

The question of the number of discovered sites is a highly important one for the fieldwalker, for the team is paid on the basis of the number of sites found (on the assumption that the effort expended and time taken to conduct the search is greater if more discoveries are made). The definition of the boundaries of sites, and thus their number, becomes for the fieldworker not only a theoretical problem, but an ethical dilemma with a financial aspect.

Definition of the boundary of a site in the context of its legal protection

The problem of boundary definition of archaeological sites is of no mean importance, for if one wishes to protect the site by scheduling it (including it on a regional or national register of protected sites), one has to be able to define the area protected clearly and in a scientifically justifiable way. In cases where this action restricts the rights of the owners to use their property as they wish, one has to be careful to define the extent of the area that has to be treated differently from the rest, and to be able to defend such a decision in court if the need arises. In such cases, however, archaeology seems less than fit to deal with the situation. Intuitive line-drawing, even if backed by scientific standing, often has little status in a court of law.

In the case of sites discovered by the AZP and only known on this basis, the definition of their boundary for the purpose of issuing an administrative decision protecting them by law relies on the distribution of surface traces. This need not always represent the extent of the site under the ploughsoil. Part of a ploughed site may for example be covered by deep hillwash and undetected by fieldwalking. The Roman period settlement at Kryspinów near Cracow, for example (excavations directed by P. Kaczanowski of Cracow University), was scheduled on the basis of the evidence of fieldwalking during the AZP project, but, when the owner wanted to build on it, it transpired that the actual extent of the settlement under deep and sterile hillwash was much greater than anticipated, and, since that part of the site had been omitted in the scheduling, the subsequent excavation had to be funded by the state and not the developer.

In the case of the Mesolithic site at Dobra near Szczecin, a classic (textbook) example that was threatened by housing development, the developer contested the decision of the local monuments curators on the grounds of their inability to define the precise boundaries of the site on the basis of the excavated evidence (this sand-dune site adjacent to peaty wetlands has for the past few decades at least been under grass). The long legal battle over this site threatened to make it a precedental test case. Fortunately, on 24 February 1998 the highest administrative court of Poland decreed in favour of the conservation services. The case did not set a dangerous precedent, but unfortunately the site was lost to science, because the developer built a house in the middle of the site and dug a lake in another part of it in the course of the long legal battle. This case more than any other has brought home to the Polish archaeological community the importance of the question of defining the edges of sites that it intends to protect.

These problems would be resolved by the acceptance of the concept of conservation zones instead of that of a site. Such a concept exists, for example, in Scandinavian archaeology (B. Wyszomirska-Werbart, personal communication). A conservation zone could be declared around the site and part of its surroundings, which would resolve the problem of the difficulty of defining the precise 'edge' of a site. The boundaries of a conservation zone could be arbitrarily drawn, for example, along modern boundaries. Unfortunately, this concept will remain just that until the legal framework underpinning the protection of the Polish archaeological heritage is changed.

Multicultural (multicomponent) sites

Another problem concerning the definition of sites is the relatively frequent occurrence of material of different dates in the same restricted area of the landscape. A settlement of the Iron Age may produce a few Neolithic pits and a few flint points of Mesolithic type. Do we then have one site or three (a Mesolithic camp or kill-site, a Neolithic settlement and an Iron Age settlement)? In terms of our definition, the area where these three types of evidence occur together and the blank area surrounding them where this material does not occur are 'a site', and features of different periods may intercut. On the other hand, the spatial distribution of the material of varied periods will differ unless the boundaries are defined by some landscape features; the Neolithic pits may cover only part of the area of the Iron Age village and extend beyond the edge of the Iron Age features. The Mesolithic flint scatter may occupy only a few tens of square metres of the later scatter of features. One site or three?

Urban sites

The KESA cards are primarily orientated to recording ploughsoil sites, but there is, however, a difficulty in the adequate recording of earthwork sites. Another difficulty has been the extending of the system into the area of historic towns (Jaskanis 1989–90), where the archaeological deposits are usually concealed under buildings, paved surfaces and grassed areas and are only revealed by individual exposures (service trenches and other disturbances as well as archaeological trenches). This introduces terminological questions, such as what is an archaeological 'site' in the concept of an urban unit (the whole unit, its component parts or perhaps individual exposures of the archaeological fabric of the town?). If the latter system is used, how far beyond the boundaries of the ancient built-up area (or perhaps that of the 19th century) should this extend (for example, the mediaeval town ditch and perhaps external cemeteries should not be separated from the town walls).

Underwater sites

Another problem is the use of the card in the recording of archaeological finds under water (e.g., wrecks or inundated settlements). Such sites cannot be recorded adequately on the Archaeological Site Record Card (KESA). The Office of the Conservator-General has recently begun cooperating with the Central Maritime Museum in Gdańsk with the aim of producing a version of the archaeological site evidence card for maritime and inland underwater sites.

Upper chronological limit

A final subject of discussion that should be mentioned here is that of the upper chronological limit of the survey. In Poland, post-mediaeval archaeology has not always had the popularity it has achieved elsewhere in Europe in recent decades, and at the beginning of the project there were very few archaeologists who were willing to take scatters of post-mediaeval pottery seriously in ploughwalking surveys. Nevertheless the upper chronological limit of the survey was initially set at the 18th century (the period of the Partitions of Poland). In the 1984 instructions, however, this time range was extended to the 19th century inclusive. This marked a break with the emphasis of the AZP project as a tool for investigating prehistoric and other ancient settlement and a move to a landscape-history based project. In some regions of Poland, especially in the mountains and some wetlands, there is little evidence of settlement before the Post-Mediaeval period. The depiction of the scattered finds of Post-Mediaeval date around existing villages has been a problem. According to the 'Instructions', these widely distributed Post-Mediaeval finds 'occurring over wide areas of fields around an existing village can be marked as zones without defining the boundaries exactly' (Konopka 1984a: 10). The method of showing these zones has not been specified.

Methods of conducting the survey

The fieldwork of the AZP project takes place within the framework of rectangular search modules 5 × 7.5 km (a sheet of the 1:25,000 base-map of A4 size; Figure 7.4). The country is divided into some 8000 of these (Figure 7.5). Each team (usually 3–7 people) is assigned one of these by the local monuments curator. The aim is to record each site that is to be found in such an area on a standardized record card (the KESA). Aspects of the methodology and the discussion surrounding it have been published in two books, Konopka (1981c) and Mazurowski (1980). The book edited by D. Jaskanis (1996) summarizes discussion resulting from the first 16 years of operation of the project.

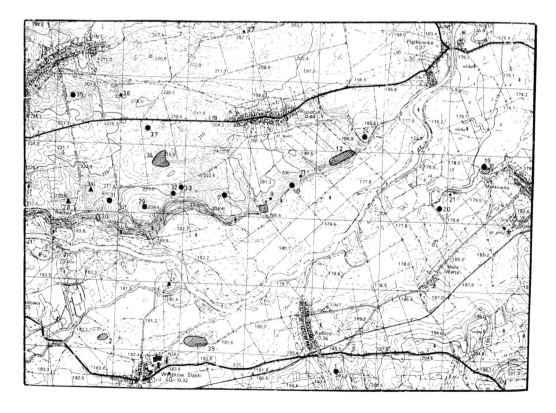

Figure 7.4   Example of the AZP module map with archaeological sites marked (Krawczyk and Romiński 1994).

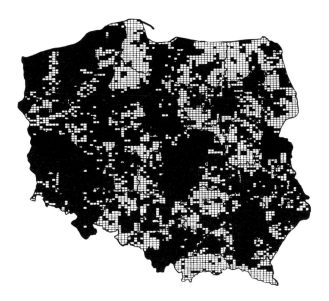

Figure 7.5   The AZP national grid with searched modules marked black (situation for 31 December 1997, according to the data provided by the ODZ).

The programme was to be realized by all archaeologists working in Poland, regardless of their place of employment. The fieldwork was to be carried out in a standardized manner, and to aid this, official 'instructions' (latest edition Konopka 1984a) were produced and a series of training sessions were run, the completion of which gave the right to lead field teams. The fieldworkers should ideally be archaeologists with some experience of working in the specific field conditions of the studied regions and familiar with the range of artefactual material likely to be recovered.

The fieldwalking element of the programme was intended to recover all archaeological sites within each search module, and the resultant picture was thus intended to reveal an 'objective' picture of the location of settlements and to produce a picture of the dynamics of settlement processes (in order to obtain such a picture it was important therefore not just to reveal where sites were present, but also to provide information on areas where such sites were absent).

Although in principle the project was carried out to a standardized methodology, in practice some variation may be observed. Some of the search modules that were investigated in the early years of the project (Koj 1996) were done to a lower standard than those of later years that drew on experience gained. Not only the quality of the fieldwork has increased, but also the quality of the presentation of the results, the latter being important not only from the point of view of aesthetics, but also in cases when the basic documentation may have to be used outside of the closed ranks of fellow archaeologists (for example, in

court to convince an appeal judge of the need to protect a particular site from the intentions of the landowner). It is particularly important that the standardized cartographic symbols and manners of filling in the KESA record cards set out in the original 'instructions' (Konopka 1984a) are firmly adhered to in order to avoid any ambiguity.

The first stage of work before the team goes into the field is that of conducting a thorough literature and archival search, revealing the state of existing knowledge about this area. This search should not rely solely on the archives of the regional services, but should reveal data that are intended to supplement them. This includes a search of museum and non-museum collections to find out what archaeological material has already been collected from the search area (in order to facilitate this, ODZ undertook the compilation of catalogues of the unpublished material lying in the storerooms of various archaeological institutions, such as universities). An attempt is later made by the AZP field teams to verify in the field any sites previously known. Those that cannot be located on the ground are recorded as 'archival' sites.

After completing the initial archival assessment, the team conducts the detailed examination of the terrain field by field, collecting material and making notes on the sites, leaving the compilation of the final record cards until later. When a site is identified, a 'statistically valid sample' of material is collected from its surface. The fieldwork takes place in a few weeks in spring and is continued in autumn, when the ground visibility is optimum. Fieldwalking consists of the team forming lines and moving across the landscape together. Mazurowski (1980: 54–5) suggests that each team member occupies a strip about 30 m across and walks in a zigzag within it, covering about 2.5 km an hour. If the strips are spaced 50–100 m apart, the average team can cover about 0.7 sq km in an hour. While accepting that information may be lost by such a method, he regards more closer spacing as not cost-effective.

In the communist period in Poland concepts of private ownership included the possibility of the use of private land for public purposes. Few farmers paid much attention to small groups of people walking all over their land (as long as they caused no great damage). Most of the small farmers had received land from the state in the land reforms, and ownership rights were understood somewhat differently than in western Europe. Few fieldworkers relish the prospect of having to obtain permission to conduct fieldwork in individual fields, because the question of the precise present ownership of many pieces of land is sometimes still an open one, and tracing the owners of the several hundred small strip fields into which some areas of the country are divided (while in theory possible) can sometimes be a daunting and extremely time-consuming task. Since the majority of the farmers can be counted on still to hold the old attitudes, most fieldworkers at present understandably tend to avoid the problem. This situation only causes problems when attempts are made to protect sites discovered by the AZP. The lack of formal consent by the landowner to conduct fieldwork has recently been used by courts to overthrow administrative decisions on the legal protection of some sites. This is a dangerous precedent, for it means that a landowner may, by withholding permission to conduct fieldwork, prevent the imposition of the restrictions on the use of his land that would apply if it was a legally protected ancient monument.

The information from the AZP survey and archival work was to be held on the KESA, which is the basic form of the archive to which other elements can be added as needs dictate. This card is to be completed in at least two copies (one for the regional conservation services, the other for the central archive in Warsaw). The record card, which is in the form of a questionnaire, is arranged in such a manner as to provide a common scheme of recording of all sites, whether known from archives alone or located in fieldwalking. The card is constructed so as to require the minimum of writing, many observations being noted by simply placing a cross in the right box. While this simplifies the recording and speeds eventual computerization, it does tend to force the investigator to make sometimes arbitrary decisions with no space to justify them. The layout of the card has already been published in detail with an explanation of the significance of the various information fields (Jaskanis 1987; 1992). In brief, the front of the card has boxes referring to the name and location of the site, its topographical location (macro- and microscale), accessibility, chronological classification and function (inasmuch as these may be determined from the surface material), nature of threats, recommendations for the conservation services and place for the signatures of those verifying the records. On the reverse is a space for a 1:10,000 location map and further remarks.

Attention should be drawn to several facts. It is obvious that a record card of this type is essential for the standardization of the recording of information. The empty fields, each of which has to be filled, require a sequence of questions to be put to the data in the field. These cover such aspects of the site as its topographic position, soil type, visibility and character. The need to compile a universal record card that is applicable to all types of sites in all topographical settings has, however — it should be said — led to the reduction of the systematized information to its lowest common denominator. The space on the reverse of the card for a free comment supplying additional data, is the field where the

observational and cognitive abilities of individual fieldworkers become most visible.

The KESA record card is primarily designed for the recording of one type of manifestation of archaeological data in the field, scatters of artefacts in ploughsoil, which constitute a major category of site in Poland's archaeological record. Another class of site that appears frequently in the AZP survey is the earthwork monument, such as burial mounds and strongholds. The AZP card should be supplemented by detailed contour surveys of these earthworks (in contrast to, for example, the English school of landscape archaeology, minor earthworks tend not to be noted, and are rarely the subject of AZP records, even though they may be reflections of significant episodes of landscape history). Standing architectural monuments, such as town houses, surviving rural architecture and churches, are as a rule not to be included in the AZP records (but are listed as such elsewhere in the archives of the ODZ).

One very serious difficulty with the current form of the AZP project is the inadequacy of the original map base. When the scheme was set up, in the specific conditions of the late-communist state, the only maps available were already outdated 1:25,000 maps (Figure 7.4). The main difficulty was the sociopolitical situation, characterized by a pathological Cold War compulsion for making a state secret out of basic topographical data. These maps are now an extremely inadequate representation of the current situation, not only with regard to anthropogenic elements of the landscape (buildings and roads and of course military installations), but also natural features, such as forests and watercourses. If this was not enough, these maps contained deliberate distortions of the position of certain landscape elements (including the insertion of triangular areas of non-existent countryside to distort the grid, so as to make its use impossible for tactical targeting of nuclear missiles). Another ploy of military origin was to exaggerate the steepness of some slopes. In communist Poland all large-scale (above 1:100,000) maps were supposed to be treated as secret documents even in the 1970s and 1980s. The possession of such maps was regarded as a privilege and the institute in receipt of such maps was obliged to keep them locked when not in use and in a safe overnight. The very act of walking across the countryside with a map and taking notes could be regarded with suspicion. In the early phases of the Cold War these restrictions were probably treated by the authorities more seriously than by archaeologists in Poland in the 1970s and 1980s, and there were no restrictions on the publication of fragments of maps in archaeological publications (which were not in any case subject to the attentions of the censor).

Another problem was the lack of a national grid on the original sheets, which forced the investigators to use an unsatisfactory system of location by measurements within the search modules. This would have been acceptable were it not for certain inaccuracies that crept into the location of the edges of adjacent modules with respect to each other. The boundaries of the modules were originally determined by the central coordinator in conjunction with geographers from Warsaw University in 1978. As it happens, when seen in the field, some of them overlapped for some reason, and some were separated on the ground by several tens of metres. These differences were not always regular in shape, and in part some were caused by the deliberate distortions in the base maps. This caused problems in the definition of the location of sites falling on the boundary between two sheets. In the late 1980s, 1:25,000 maps with a national grid became generally available and the long process of translation of the original measurements into national grid references could be begun. This was of course fundamental to the computerization of the data.

Usually, an attempt is made to collect a relatively large sample of the material from each site, but not all of it. The washed and marked material is examined by the consultants, who check the identification of the material, and then packed and deposited in a museum collection as part of the AZP archive. The collection's inventory number is entered on the KESA card. The material is therefore available for later examination. The recognition of the cultural affinities and chronology of a handful of small eroded sherds found loose on the surface of a ploughed field often causes problems (Czerniak 1996: 39–40). Some investigators and their consultants showed greater imagination or confidence in such attributions than others.

Some later investigators building on the AZP results rely on the original identifications, while other prefer the more time-consuming chore of returning to the original material in museum collections. Sometimes this exercise in source-criticism can produce results which differ from the original ones. As an example we may quote Professor M. Parczewski's re-examination of the evidence for Early Mediaeval settlement in the extreme southeast corner of Poland (Parczewski 1991: 18–19). He not only verified the Early Mediaeval sites, but also examined the 'prehistoric' pottery, among which he found sherds he had little trouble in accepting as Early Mediaeval, which affected the overall picture somewhat. Professor W. Szymański examined 'Early Mediaeval' pottery from the AZP collection from Plock Province and found that most of it should actually be dated to the Roman Period. In another case Professor J. Dąbrowski examined all the supposed Lusatian Culture (LBA-EIA) pottery from the northeast

corner of Poland and found that about 50% of these assemblages had been incorrectly identified (D. Jaskanis, personal communication). Bienia and Żółkowski (1996: 153–55) cite a similar 50% error rate. It should be noted, however, that the results of the culture-chronological identification of the same sherds done by various scholars can sometimes differ quite dramatically. Another problem has been with changes in culture names and attributions of affinities over the past few years. Some new cultures have been defined, others renamed and amalgamated.

As we have seen, the AZP project in general relies on stripwalking to define the location and boundaries of a site. Collection of artefacts within the site according to a grid is rarely practised as part of the 'first stage' of the AZP survey. The discussion concerning the so-called 'second phase' of the AZP survey (Jaskanis 1996), includes, however, the possibility of conducting detailed planigraphy of the distribution of artefactual material across the surface of the site. One of the first attempts to define and analyse concentrations of artefacts on the surface was presented by Rózycki in 1980, using sophisticated mathematics. A recent paper (Bitner-Wróblewska *et al.* 1996) considers simple ways of dealing with this problem, but the authors seem to forget that the distribution of surface finds need have little connection with their original position within a site (also one must note that, when seen in the context of the European literature as a whole the authors are in error in referring to this as a 'new' technique).

Such problems, however, only emphasize that the AZP is an ongoing project, and cannot be regarded as 'finished'; they do not detract from the value of the project itself. From its inception, the AZP project was considered in some quarters as a never-ending programme and cannot be regarded as completed when each search module has been walked just once (such an approach would seem to be wrongly suggested by Konopka [1984a: 9], where the discovery of 10% new sites in a second fieldwalking programme would disqualify the search module as badly done).

The AZP as a tool for heritage management

Copies of the results of the AZP records with explanatory notes are sent by the provincial inspectors to local councils for use in monuments protection during the planning process and the delimitation of conservation zones in local planning documents. Any redevelopment within these zones (of which there are several grades) must be agreed by the regional inspector of monuments, without whose acceptance the work is considered illegal. This creates a mechanism by which the results of the AZP project become the basis of regional heritage management plans.

The conservation needs of the sites revealed by AZP survey was one of the key features of the original project design. In the *Principles of the Realization of the Archaeological Field Record of Poland* (a document issued by the Conservator-General, Professor W. Zin on 15 February 1980), considerable space was given to the uses of the AZP project as a tool for the conservation of the archaeological heritage: '[T]he results of the investigations and their documentation should be analysed by the provincial archaeological inspectors and used in the planning of immediate and long-term activities connected with archaeological investigations and conservation work.' It should be noted here that in this document of 1980, the Conservator-General instituted a policy that the conservation services are still having problems getting the academic community to adjust to, that is, the policy that only threatened sites should be the subject of excavations, be they rescue or research excavations. This is in agreement with the principles laid down in later international conventions. The document then goes on to say:

> 'These results should be analysed from the point of view of
> a) identifying sites for inventorising in the second phase of the project (surveying, detailed surface collection, trial-trenches [...] geophysical investigation, aerial photography etc.),
> b) identifying sites for inclusion in the list of scheduled sites, for the erection of information tablets, the organisation of monuments wardens for them, or the establishment of archaeological conservation areas,
> c) selection of the most threatened sites for rescue and research excavations.'

In the official 'instructions' concerning the relevant portion of the front of the KESA card, Konopka (1984a: 35) warned against 'devaluing the value of the information that a "threat exists", since in each case this information should prompt a rapid action by the conservation services'. This was to take the form of a field inspection followed, if necessary, by administrative action or archaeological excavation organized by the archaeological monuments inspector himself, or by 'interesting an archaeological institution' in the site. Unfortunately, what an 'important threat' was (Konopka 1984a: 34) was not defined, nor what was an 'unimportant' one. The AZP field team is also expected to define the cognitive potential of each site. The instructions do not, however, contain any clear guidelines as to how the fieldworker should define this (quantity and density of finds? typicality or

atypicality? existence of one phase only? or alternatively, existence of several phases? rarity of sites of this type/period in the region? potential existence of organic remains?). The 'instructions' suggest that this assessment should be made with the help of the consultants, but does this mean that they would have to take part in the original fieldwork? or is a 'desktop' assessment meant?

It is not clear, seen in a 20-year prospect, what the practical results of this process have been. Information on the number of archaeological sites that have been protected by law have been collated by D. Jaskanis (forthcoming); it seems, however, that they form a very small percentage of the 300,000 sites known to date from the AZP project. Clearly, in the face of the scale of current threats (Kobyliński 1997), the future growth of Polish archaeology in succeeding decades should be based around a coherent heritage management scheme, including elements of a project similar to the British Monuments Protection Programme (English Heritage Archaeology Division 1992). It remains to be seen to what extent the AZP records will be able to serve as a basis for such a scheme. One problem that has yet to be addressed in Polish archaeology is that of plough damage, which is of course what brings to the surface the material we collect as archaeological data in the AZP project. There is a direct relationship between the uncontrolled destruction of the evidence and its recognition. Until recently in many parts of the country the principal form of rural traction was the horse, and horse-drawn ploughing was a common sight in the countryside, even near the capital. With the present economic changes, heavy farm machinery is replacing the horse and allowing deeper ploughing of archaeological sites. We urgently need a survey of the scale of the problem, its effects and possible ways of counteracting it (see Hinchliffe and Schadla-Hall 1980).

A major difficulty in the use of the AZP project as a heritage management tool is the surface non-detectability of some sites. If a site is not found in the AZP survey, it is not included in a conservation zone and thus does not appear in the desktop assessments in environmental impact surveys in the planning process (indeed the documentation of building plots outside the conservation zones marked on the planning department's policy map is seldom even seen by the regional monuments inspector or his staff). A similar problem has appeared in the projecting of a motorway system in Poland, where financing has been set aside for the excavation of sites known from past AZP work as well as new surface surveys conducted along the routes after their planning. Sites that were undetected in both passes will have to be recorded during the earth-moving of the development, when there may not be adequate legal and financial provision for their proper investigation.

As yet there seems to be a complete lack of policy on this matter in the organizations responsible for the initiation of rescue archaeology projects on the motorways, which continue to be a source of worry for the state conservation services.

The AZP as a tool for public education

Although this was the third of the stated initial aims of the AZP project, it is also the one which has not been fulfilled. The whole aspect of the interface of the AZP project with the public seems to have been somewhat neglected (Matoga 1996: 54). It has to be admitted that the project in its present form has made very little contribution to public knowledge and appreciation of the early history of Poland. One has the impression that rather too much time has been given to discussions on the collection and storage of information coming from the AZP project, and rather too little to the methods of making them available to the public. This tends to ignore the fact that this very expensive exercise is funded almost entirely from the public purse. It is an expenditure of which the general public is on the whole completely unaware. If we are to escape the inevitable political consequences in a democracy, of the continued ignoring in this manner of the existence of the general public, one of the principal aims of the project and the conservation services in general must be to change this situation, working with schools and the media to propagate the ideals of the conservation and responsible management of the diminishing archaeological resources of our country.

One other way in which this could be done is to involve local communities in the actual conducting of the fieldwork. British experience shows that local inhabitants can be very quickly taught to differentiate pottery from flat stones and asbestos roof tile, and after initial training become just as useful in the field as first- and second-year archaeology students. The involvement of amateur archaeologists in Polish archaeological fieldwork is for various reasons an unexplored avenue.

In Poznań Province, the conservation services have produced booklets detailing the archaeological sites of particular, lowest-level local government administrative units (*gmina*). These publications (the initiative of Doctor Andrzej Prinke) are actively promoted within the community, and seem to have had some effect in increasing awareness of the problems of the protection of the cultural heritage in the areas concerned.

Some problems in the execution of the project in 1996 and 1997

An important problem connected with the execution of the AZP project is the verification of its results.[2]

At present, this has three stages. In the first verification the regional monuments curator examines all the record cards and has the possibility of checking if the archival research has been done to a high enough standard. As an archaeologist who knows the territory under his control well, the curator should be in a position to assess whether or not the location of sites has been correctly indicated, and to examine the manner in which sites have been differentiated. After the curator accepts and signs the cards, qualifying them for further verification, they are sent to the Council of the AZP project. At this stage in the verification of the documentation, the Council checks whether it has been done in accordance with the 'instructions' and complies with the national standards. At this stage the differences between the documentation from different provinces become more visible. The most important cause of this are the gaps in the 'instructions', the last revision of which appeared in 1984 (this has highlighted the urgency of the revision of this perhaps too brief and rather too vague document, which is currently being undertaken in the light of further experience and problems). The main problems concern the manner of completing the reverse of the record card, the use of a precise 1:10,000 scale map to show the location of the site, the types of cartographic symbol used to denote specific types of site, the inclusion of the correct designation of the map, etc. During the 20 years of the functioning of the AZP project, and despite attempts towards standardization, several local 'traditions' of completing the record cards have arisen, which have a negative effect on the maintenance of national standards. This has created additional problems in the computerization of the results. In the creation of these local traditions, both the fieldworker and the local monuments curator point to the lack of a definitive statement on a number of subjects in the official 'instructions', and disregard the possibility of enlarging the scope of the information recorded as a free comment on the reverse of the record card in the space reserved for this.[3]

In some provinces, in areas containing a specific form of archaeological site (for example, in Kielce Province, where there are complexes of sites connected with the ancient smelting of iron), the use of 1:25,000 maps has been abandoned because the frequency of sites on the ground would result in too dense clustering of points in certain areas, leading to illegibility of the map. These sites, sometimes of considerable size, are at once plotted on a map of scale 1:10,000.

Another problem concerning the execution of the AZP project is the penetration of areas of limited accessibility. This concerns, for example, the areas in the northeast of Poland, which are to a large extent covered in forests, meadows or wasteland.

The penetration of the forests, which may contain barrow cemeteries or strongholds, then has special significance. In such cases significant help may be provided by examination of maps of these areas kept by forestry commissions and conversations with foresters. On the cartographic documentation of AZP search modules, these areas should be clearly marked.

Practice has shown that not all fieldworkers paid sufficient attention to the conducting of the archival research preceding the fieldwork. The archival research and the verification of previously known sites in the field are among the most important aims of the programme. Examples have been found of fieldwork that has been conducted with the omission of such a procedure. In such cases, the investigation of these areas cannot be said to have been completed, and their results cannot be included in the central archive in ODZ.

The AZP project has shown up a very important and very serious lack in the SMRs of the archaeological monuments conservators in many regions of Poland. It transpires that until the institution of the AZP project, insufficient attention had been paid in several areas to the inventorization of archaeological discoveries made before the Second World War. This particularly concerned the so-called 'Recovered Territories' in the north and west of Poland, formerly parts of German territory that were assigned to Poland by the Teheran and Potsdam Conferences. These territories had been investigated by German archaeologists since the end of the 18th century. After the war, however, the low staffing levels of the conservation services and the scale of destruction caused by the war and redevelopment of the region created a situation where the conservation services were not in a state — despite calls to this effect (Kostrzewski 1950) — to assimilate the information from these old finds (much of the archival material had been destroyed or muddled in the war, many museum storerooms, archives and libraries had been evacuated, looted and shelled). The old archives from these areas that have survived in Poland are in handwritten Gothic script, in German, and are often difficult to use. In recent months the Office of the Conservator-General has succeeded in gaining access — thanks to Doctor Judith Oexle, archaeological curator for Saxony — to a small part of the German archival material referring to old discoveries from Poland, which had been kept in Görlitz.

For this reason, only the assimilation of the old data into the AZP records would create the basis of an archive in some of these provinces (Jaskanis 1996: 10). A paradoxical situation was thus created. Instead of the fieldworkers being able to draw on a full list of sites compiled by the conservation services from old publications and museum records, AZP fieldworkers were actually responsible for creating this basic

record, search module by search module, which led to a repetitive full literature search of dozens of old publications including 19th-century periodicals for each area. This seems a rather 'backward' way of compiling a basic heritage management tool. In areas where the conservation services do not possess these data, there is also no measure against which the achievements of the compilers of the AZP record can be assessed. This would seem to be one of the weakest links in the present form of the AZP project.

Among the most common mistakes made by the fieldworkers in compiling AZP documentation is the inconsistent use of cartographic symbols for different types of sites, confusing the concept of 'trace site' and 'settlement point', the use of additional symbols (without supplying a key), for example, shading areas of the site, and lack of inventorying of the finds. Sometimes one finds careless completion of the record cards, the omission of some information in specific fields or the numbering of sites on the module map by hand without the use of stencils or letraset.

The third stage of verification concerns the correctness of the chronological and cultural identification of the finds collected during the fieldwork. This is done by groups of consultants — archaeologists of long experience working in the area — responsible for each province. Together with the regional monuments curator they conduct a detailed examination of the material collected by the fieldworkers and correct the assignations of finds to specific cultural groups and chronological ranges. Sometimes this process of verification of the finds takes place during the preparation of the final documentation of each search module; this is the case in certain parts of southern Poland. In most cases, however, the consultants meet to examine the material from several areas (thus allowing them the possibility of comparing finds across a wider area) after the records have been accepted and paid for by the regional monuments conservator. In such cases, any corrections have to be made on the finished cards and are not included in their contents. The need to check the cultural and chronological assignation of finds derives from the enormous variety recovered by fieldwork. In many areas the chronological range of these finds can extend from the Palaeolithic to almost the modern period. From information in the possession of the Council of the project it seems that the verification carried out by the consultants has sometimes significantly changed the interpretation of the results. The original identifications, which are often made tentatively in the field by the fieldworkers, often require modification. This is an important problem in the nature of source creation because, if the fieldworkers are unable to understand what they are collecting, then they cannot be collecting all information about its context (note that they are expected to conduct some form of 'selection' of material in the field); for example, in not being able to differentiate Palaeolithic from Neolithic flints, they may be missing important information about its zoning within an extensive site. This fact has significance in the debate surrounding the eventual publication of the results of the AZP project, which, regardless of its form (see below) must include the results of a detailed examination of the cultural and chronological verification of the finds.

The AZP documentation is also additionally checked from the formal and meritorial aspects by the Archaeology Department of ODZ. This institution is the destination of the copies of the record from the whole country, forming a traditional paper archive from which a computerized database is prepared.

The checking of the results of AZP surveys is mainly confined to the documentation concerning the work, and verifies the correctness of the chronologcal-cultural identification of the finds. The verification of the fieldwork on newly-discovered sites involves many more difficulties. The 'instructions' foresaw this as a matter of re-examining certain areas, and postulated the disqualification of the results of the first search when there is a difference of 10% in the results between first and second searches (Konopka 1984a). In general, in the assessment of the documentation, it is assumed that the fieldwork was correctly carried out. This is not always justified, as is made clear by reports from the first years of AZP surveys in various areas (such as, for example, Częstochowa Province; Koj 1996). It should be remembered, however, that, as discussed above, a second pass over the area, especially when done sometime later and in different weather conditions, etc., may lead to the discovery of new sites and the non-discovery of other sites previously visible (Czerniak 1996).

The AZP and other forms of prospection

The above-mentioned restrictions on the cognitive values of the AZP project mean that in Polish archaeology there is a discussion concerning the so-called 'second stage' of the AZP project, which was to form a supplement to the fieldwalking. In the opinion of several archaeologists this should take the form of trial-trenches (Matoga 1996), but, as is well-known, the cognitive value of a trial-trench may be just as restricted as fieldwalking. On sites consisting of isolated negative features but with no horizontal stratigraphy below the ploughsoil, it may be necessary to open large areas before the nature of the site becomes clear. Small trenches, shovel-probes or augering — no matter how numerous or closely spaced — may

give quite a false picture, such as resulted from a survey preceding the gas pipeline construction in eastern Poland (D. Krasnodębski, personal communication). Other methods of prospection, such as phosphate analyses or geophysical prospection, may supplement the surface artefact data.

In the early 1980s in Poland the first attempts were made to apply representative sampling methods in the investigation of settlement sites, and especially non-invasive methods for the examination of the cultural layers and the situation of archaeological features. This was carried out, for example, in the case of a settlement from the beginning of the Early Mediaeval period at Wyszogród-Drwały in Płock Province, where random sampling was applied to a site, the extent of which was already known from surface fieldwork. Samples were taken by augering at randomly chosen points on a grid, the size of which was fixed relative to the average size of features uncovered in previous trial-excavations. The results of these samples were then checked by excavation. In the places where the presence of culture-layers had been detected, archaeological features were discovered. The use of this method considerably facilitated the understanding of the settlement in a labour-saving and cost-effective manner. It was discovered that, although the area and extent of the site had been defined by fieldwalking, the area of the actual preserved occurrence of the culture layers was considerably smaller and had a discontinuous character (Kobyliński *et al.* 1984; Brzeziński *et al.* 1985). A modified version of this method was applied to define the extent of the occurrence of the cultural layer on a settlement at Rostek, in Suwałki Province (Brzeziński 1990).

The occurrence of phosphates in soil can also be used in archaeological settlement studies not only to determine the intensity of human activity on a given site, and also to designate the zones within a site and the manner of their use, but primarily to define the extent of the site itself, manifested by the appearance of artefactual material (Brzeziński *et al.* 1983).

Geophysical prospection methods, such as resistivity and electromagnetic methods of exploration, can form an important supplement to investigations that define the extent of archaeological sites by non-invasive methods. These have been used in Poland in tracing the extent of flint mines (e.g. at Krzemionki), where they were used to examine the extent of the mining shafts and the extent of the underground workings without excavation, which would be impossible in the case of a site this size (e.g. Herbich 1993).

A method that has been found to be very useful in other European countries is aerial photography. Although Polish archaeologists have long been aware of the value of this technique (e.g. Gąssowski 1983: 175–215), it has not been used in Poland to the same extent as in some other countries. This was primarily due to the same factors involved in the secrecy generated around cartography, but also economic factors and lack of easy access to suitable aeroplanes played a role. This has until now prevented the supplementation of the fieldwalking data by information from aerial photography. Exceptions to this general rule were the work done by Ewa Banasiewicz in the late 1980s and 1990s in Zamość Province, and Włodzimierz Rączkowski in Słupsk Province. In 1996 the office of the Conservator-General initiated a series of test flights in selected regions of the country, facilitated by the help of experienced foreign flyers, such as Otto Braasch from Germany and Martin Gojda from the Czech Republic. The good initial results of these flights led to an increase in funding for this form of prospection in 1997, when prospection was carried out in 17 provinces. It is now clear that a number of sites that produce pottery scatters also produce cropmarks, while some sites appear only as cropmarks. This form of survey should in future form one of the main tasks of the second stage of the AZP. Incorporation of the aerial photographic data with the AZP fieldwalking data will probably have a considerable impact.

Computerization

The degree of computerisation of the AZP record is somewhat unsatisfactory. Despite the long running of the project, this vital aspect has lagged behind. The computerization of the AZP records had its beginnings in the programs of the mathematician and programmer Bogdan Gliniecki and the archaeologist Andrzej Prinke. These initial programmes were revised many times. Beginning in dBase Clipper, later converted to FoxPro as 'AZPFox', this was a relatively simple data-management programme (Prinke 1996). This is currently in use in the central ODZ archive as well as those of the regional archaeological inspectors. Up to now only 30% of the information has been computerized, and to varying degrees in different provinces (Jaskanis 1996: 37). A fault of this system was the impossibility of printing the record card in its standard form, and of that of connecting texts from the database with illustrations of the finds, or with maps. Such a possibility appeared together with MS Windows and GIS programs. The Office of the Conservator-General has since 1996 postulated the conversion of Prinke's programme to Windows, but this was preceded by a local initiative of the regional monuments curator in Wałbrzych in southwest Poland working with the regional planning office. Here Piotr Łapkowski and Marek Kowalski connected a database created in MS Access with the MapInfo program, and in October 1996 were ready to present the working version showing sites discovered

in AZP surveys on maps of various scales, on which could be overlaid data from vertical aerial photographs and other information concerning for example planning zones. After the demonstration of this program, Prinke, with financing from the Ministry of Culture and Arts (Office of the Conservator-General) and the Polish Committee for Scientific Research, produced in 1997 a new version of his program based on MapInfo. The version currently available unites the AZP database with the possibility of drawing maps of various scale showing the distribution of archaeological sites. This creates totally new possibilities for the utilization of the results of the AZP project in scientific publications, but above all in heritage management, as a planning tool and in the preparation of desktop environmental impact assessments, etc. The value of both systems is that the information can be stored and reproduced in a format corresponding to the layout of the original cards. Another interesting feature is the possibility of overlaying vertical aerial photographs with data from maps or the SMR.

Independently of these programs in several provinces (e.g. Toruń, Sieradz, Kielce, Piotrków Trybunalski) the local monument curators, not satisfied with the software then available from ODZ, designed their own, relatively simple programs based on commercial packages in order to complete and print AZP site record cards.

Publication

The accumulation of such large quantities of data has raised questions of what to do with them? who is their owner? and who should have access to them? The question of copyright to the information contained in the documentation of AZP fieldwork is a problem that has yet to be satisfactorily resolved. While the project is state-funded and carried out to a predetermined methodology, individual fieldworkers feel that they have some form of rights to their discoveries. A rather odd situation thus develops, whereby material that is held in a public archive is to be treated as held in some form of trust for the original discoverers, and yet the original 1978 statute of ODZ stated that the archive was to be made wholly available to the public, except (typically) if releasing certain types of information was against the interests of state security. It hardly seems that the distribution of a few handfuls of sherds in muddy fields falls into this latter category.

Although from the very beginning of the AZP project, some form of standardized publication of the results of the whole programme on a national scale was envisaged, no agreement was reached on this. This has meant that the results achieved until now have been unpublished or have been published in a variety of forms. Some investigators (e.g. Tomczak

1994) think that the results of the AZP project should be published *in extenso* in the form of site lists and detailed maps containing the exact location all sites. Others regard this as undesirable from a number of viewpoints. The first is that the publication of the results would sanction the fact that the 'action' was finished in a given area after it was fieldwalked just once. Second, the AZP data are archival material available — in a controlled manner — in state archives, and publication of the data is senseless. Third, there is the danger that publications containing detailed location maps may allow the locating of specific sites by criminals, especially looters with metal detectors, who dig into the archaeological deposits and have become a specific threat to sites in Poland in the last few years. For this reason, we have decided not to make AZP data available through the Internet.

Other workers (A. Prinke) feel that the results of the AZP project should be published within the boundaries of local administrative units, giving local councils a publication that can be used to formulate planning policies and as a promotion for the aims of protection of ancient monuments.

Stefan Woyda has recently suggested that for scientific purposes one may consider publishing the distribution of sites on maps that show the topography and hydrography of an area, but omitting contemporary anthropogenic features that would allow a looter to locate the site within the modern landscape. This position is supported also by the Office of the Conservator-General. It seems that the database should be presented to the public in a 'digested' form, for example, scientific publications on the theme of the relationship between the settlement network and the natural environment.

A conference on this subject in the Office of the Conservator-General in 1997 did not allow the arrival at a consensus and this discussion must be continued.

The future

The AZP project is unique in its conception and scope. Like all long-term projects, problems have arisen in adjusting concepts arrived at 20 years ago with present needs, but practice has shown that, in the case of the AZP project — despite a certain conceptual poverty — these problems are not in fact as serious as they might have been. This is due to the relative sophistication of Polish settlement archaeology in the 1960s and 1970s, but also to the careful planning of the project, a process involving many individual scholars. In this paper we have been considering the present state of the AZP project, but the title of the present book requires us to at least consider the future of the databank thus produced.

One of the main problems which has recurred here is the question of archiving the data, in what form, and how to utilize and disseminate the information resources thus generated. The lack of nationally valid considered principles for the construction of provincial archaeological archives is one of the most damaging factors in the present state of archaeological heritage management in this country. The Office of the Conservator-General intends in the near future to initiate discussion on this subject and produce guidelines for the regional conservation services. Another major problem is integrating the AZP data with that held in the registers of standing buildings. For too long in Polish archaeology has archaeological conservation been neglected in favour of the more spectacular, obvious and visible monuments of the past. In an integrated conservation service, which is the aim of the Office of the Conservator-General, we should find a way of uniting all the information about surviving monuments, whether they be roofed, ruins or just foundations in the soil.

It is worth noting that the various editions of the 'instructions' for the scheme concerned themselves mainly with the construction of the AZP database, without going into much detail of how it was to be used. Perhaps it would be worth returning to the debate begun by the book edited by D. Jaskanis (1996) and attempt to resolve the problem of the future directions of development of the project in a period of massive changes within Polish and European archaeology. Should the methodology of the project remain static and unchanged? If we assume that it really is a never-ending task, perhaps the beginning of the third decade of its history is an appropriate time to introduce improvements that will guide its operation in the new century?

We should also stimulate further thought on the methodology and theory of surface fieldwork in Poland. The AZP project has tended to monopolize discussions on this theme, and perhaps to obscure several important issues, or at least distort our perception of them (Moore and Keene 1983). The lack of the abundant theoretical literature and summaries of experimental work that is visible in the Anglophone literature is striking.

A key question not discussed at all by the initiators and compilers of the AZP project was the order in which various areas were to be penetrated. The lack of a concept concerning this, which left it to accident or the individual preferences of the monuments curators and fieldwalkers to decide which areas were to be searched first, has led to a situation where some areas of some provinces have been surveyed in their entirety, while others have been penetrated by the AZP project to only a slight degree. Within the provinces the choice of survey area was dictated by random factors or the ease of access by car, bus or train. Because of this, the results cannot be used to draw scientific conclusions on the subject of settlement patterns or cultural phenomena before the completion of at least the first phase of the AZP project. If, however, random sampling methods had been applied, as was proposed as long ago as 1984 (Kobyliński 1984: 12), it would be possible very quickly to get scientific results (if our aim was to attain statistical estimations of the resources of the archaeological heritage), and these results could have been verified by successive phases of the project.

On the other hand, from the point of view of the needs of conservation, the order in which areas are searched should not be random, but should reflect the potential threats to archaeological sites, and thus the areas around developing towns should be examined, those on the routes of planned extensions of the motorway system or other linear developments, planned reservoirs, areas destined for afforestation, and so on. From this point of view it may be suggested that areas under forest and the mountainous regions could be surveyed as a last resort (with obvious implications for the possibilities of understanding the specific forms of settlement of these areas or the recognition of previously unknown cultural units).

The AZP project, with all the theoretical limitations discussed above, can supply a large number of scientific data, which can form the basis of many important studies on the theme of settlement patterns and their environmental determinants, with perhaps reference to the well-known concept of 'site-catchment analysis' (summarized for the Polish reader by Kobyliński 1986), or the basis for the simulation of settlement processes (foreign literature on this subject presented to the Polish reader by Kobyliński and Urbańczyk 1984). There have already been published in Poland many settlement studies based on the results of the AZP project in various regions in which the first stage of the AZP project has reached its conclusion (e.g. Rydzewski 1986; Dulinicz 1990; Losiński 1982). These studies tend to concentrate on specific regions in a specific timespan and not to study such aspects as, for example, landscape development. It seems that the data held in the AZP project remain to a large extent unused as a source of information on past socio-cultural processes. There is still, as we have signalled above, a lack of critical discussion on the theoretical basis of the AZP project, which has basically meant that, in possessing this vast quantity of data, we still do not know what we have, and in what way it may (or may not) be used. If in every case we could indeed treat places where the AZP project has revealed material on the surface as a reflection of the site of ancient habitation or activity, then maps based on AZP surveys would serve for detailed geographical studies (cf. Kobyliński 1987). These maps cannot, for the reasons given above, be used uncritically in this way (cf. Dulinicz and Kobyliński 1990: 250–55), because of the impossibility of precise determination

of the chronology of the finds; the inability to determine what really lies behind the discovery of a few sherds of ancient pottery on the surface; and the dependence of observation on a variety of subjective and objective factors. All of this should mean that the distribution of sites discovered by the AZP project may be both richer or poorer than the real existing pattern. Awareness of these cognitive limitations of the AZP project is well-established among Polish archaeologists and probably it is to this that one must ascribe their unwillingness to publish syntheses based entirely on the results of surface discoveries. The unwillingness to publish serious archaeological settlement studies on the basis of the AZP project may also have two other causes. First, as was already said, the AZP project is one that cannot be regarded as ever having an end. Results of further fieldwalking, supplemented by the results of aerial photography, and new discoveries revealed by earthmoving in development all require the continuous supplementation of the AZP archives. Scholars are therefore afraid that work based on the preliminary results of AZP survey will quickly transpire to be totally false. Second, the AZP project is a cooperative exercise based entirely on trust in the professional ethical standards of those taking part. A multistage process of verification of the results can only hope to discover mistakes in filling in the KESA cards (i.e. not in accordance with the 'instructions'), or in the classification of the finds. In no way can the quality of the fieldwork be checked: whether the team working in a particular area really did exert all possible effort to locate all possible sites in the search module; whether they reached all fields in the optimum stage of cultivation and under the optimum conditions of visibility; whether they penetrated all forests and meadowland. The AZP project is thus a rare example of a real group operation, the collective labour of the whole archaeological community in Poland. As in every milieu it is possible that there are individuals who are or were less than careful in these aspects, and motivated by their own short-term interests. This is perhaps the cause of the mistrust many archaeologists feel regarding areas searched by other scholars personally unknown to them, and the tendencies towards monopolization of the conducting of AZP surveys in selected large parts of some provinces by some workers, who then are able to trust the results enough to use their own work in regional studies. This phenomenon has a positive side, for then the inspiration to conduct the survey is not a commercial one, but a scientific one.

Despite the huge potential scientific possibilities, the basic significance of the AZP project is in its usefulness in creating a SMR for use in monument protection and conservation at the stage of development planning and cultural resource management.

Here the decision on the future of the project has a particular significance. Can one regard one passage (even conducted to the highest degree of detail) across an area as adequate for determining the archaeological resources of the area and serve as the basis for their protection? We may state with certainty that this is not the case. Our attempts to finish the so-called first stage of the AZP project over the whole country is therefore an attempt to learn 'anything at all' about each area that may potentially be threatened by development, change in land use or changes in the environment, before that threat arises. The achievement of this bare minimum will not mean, however, the end of surface investigation in Poland.

As a final point, we wish to postulate that we need to propagate the aims and achievements of the project, and to generate public awareness of its value. This seems a prerequisite for its future functioning if it is to be paid for by the taxpayer.

Acknowledgements

The views presented in this article written in 1998 derive from our experience as fieldwalkers involved in the AZP project in various parts of the country, and as the national coordinators of this programme, but also as the result of many discussions with provincial monuments curators, to whom we are grateful for many remarks and suggestions. Especial thanks are due to Danuta Jaskanis, who for many years was the national coordinator of the project and who willingly shared with us her rich experience. The sterling work of Piotr Szpanowski (Office of the Conservator-General) should also be mentioned. The project would not have been such a success without his hard work behind the scenes. Figures on the financing of the AZP project and the number of areas done each year, as well as the permission to use them here, were kindly supplied by S. Żółkowski of ODZ.

Note

1) Now the 'Ministry of Culture and the National Heritage'.
2) This section presents the experience of the Council (Board) of the project from an examination of the results of the fieldwork carried out in 1996 and 1997. It draws on the opinions of the members of the Council (W. Brzeziński, D. Jaskanis and A. Prinke) copncerning the documentation of particular areas.
3) A new set of 'instructions' has now been published.

References

Banasiewicz, E.
   1996   Archeologiczne Zdjęcie Polski jako badania podstawowe. In D. Jaskanis (ed.), 1996b: 92–96.
Bienia, M., and S. Żółkowski
   1996   Weryfikacja wiarygodności wyników badań AZP w województwie bialskopodlaskim. In D. Jaskanis 1996b: 151–59.

Bintliff, J., and A.M. Snodgrass
   1988   Off-site pottery distributions: a regional and inter-regional perspective. *Current Anthropology* 29: 506–13.
Bitner-Wróblewska, A., J. Brzozowski and J. Siemaszko
   1996   Nowe możliwości wykorzystania metody planigraficznej w badaniach archeologicznych. *Archeologia Polski* 41: 7–38.
Brzeziński, W.
   1991   Badania rozpoznawcze osady w Rostek, gm. Sołdap, woj. suwalskie. *Wiadomości Archeologiczne* 51(1) (1986–90): 105–11.
Brzeziński, W., M. Dulinicz, and Z. Kobyliński
   1983   Zawartość fosforu w glebie jako wskaźnik dawnej działalności ludzkiej. *Kwartalnik Historii Kultury Materialnej* 31(3): 277–97.
Brzeziński, W., M. Dulinicz, Z. Kobyliński, B. Lichy and W.A. Moszczyński
   1985   Rozpoznawanie stanowisk osadniczych metodą reprezentacyjną: badania w Wyszogrodzie, woj. Płockie, stan. 2a. *Sprawozdania Archeologiczne* 37: 251–70.
Czerniak, L.
   1996   Archeologiczne Zdjęcie Polski — co dalej? In Jaskanis 1996b: 39–46.
Dulinicz, M.
   1990   Stan i potrzeby badań nad osadnictwem wczesnośredniowiecznym na Mazowszu (VI–XI w.). In Z. Kurnatowska (ed.), *Stan i potrzeby badań nad wczesnym średniowieczem w Polsce*, 243–61. Poznań.
Dulinicz, M., and Z. Kobyliński
   1990   Archeologiczne mapy osadnicze i ich przydatność do komputerowej analizy przestrzennej. *Archeologia Polski* 35: 241–66.
English Heritage Archaeology Division
   1992   *Monuments Protection Programme: The Evaluation and Selection of Monuments for Statutory Protection.* Typescript strategy document, March 1992.
Gąssowski, J.
   1983   *Z archeologią za Pan brat.* Warszawa.
Herbich, T.
   1993   The variations of shaft fills as the basis of the estimation of flint mine extent: a Wierzbica case study. *Archaeologia Polona* 31: 71–82.
Hilczerówna, Z.
   1967   *Dorzecze górnej i środkowej Odry od VI do początków XI wieku.* Wrocław.
Hinchliffe, J., and R.T. Schadla-Hall (eds.)
   1980   *The Past under the Plough.* Occasional Paper 3. London: Directorate of Ancient Monuments and Historic Buildings, Department of the Environment.
Horbacz, T.J. and M. Olędzki
   1994   Wyniki badań powierzchniowych na obszarze 101–48. In E. Tomczak (ed.), 1994: 148–154.
Jankuhn, H.
   1977   *Einführung in die Siedlungsarchäologie.* Berlin: De Gruyter.
Jaskanis, D.
   1987   La Carte Archéologique Polonaise: Theorie et Pratique. *Nouvelles de l'Archéologie* 28: 42–52.
   1989–90 Wybrane aspekty archeologicznej dokumentacji z badań miejscowości zurbanizowanych. *Prace i Materiały Muzeum Archeologicznego i Etnograficznego w Łodzi Seria Archeologiczna* 36: 93–103.

1992 Polish National Record of Archaeological Sites — general outline. In C. Larsen (ed.), *Sites and Monuments National Archaeological Records*: 81–87. Copenhagen.

1996a Próba oceny metody AZP na podstawie doświadczeń ogólnokrajowego koordynatora. In D. Jaskanis (ed.), 1996b: 9–38.

Jaskanis, D. (ed.)
1996b *Archeologiczne Zdjęcie Polski — metoda i doświadczenia próba oceny.* Biblioteka Muzealnictwa i Ochrony Zabytków ser. B vol. XCV. Warszawa.

Kempisty, A., J. Kruk, S. Kurnatowski, R. Mazurowski, J. Okulicz, T. Rysiewska and S. Woyda
1981 Projekt założeń metodyczno-organizacyjnych Archeologicznego Zdjęcia Polski (1975). In M. Konopka (ed.), 1981c: 22–27.

Kobylińska, U., and Z. Kobyliński
1981 Kierunki etnoarcheologicznego badania ceramiki. *Kwartalnik Historii Kultury Materialnej* 29: 43–53.

Kobyliński, Z.
1984 Problemy metody reprezentacyjnej w archeologicznych badaniach osadniczych. *Archeologia Polski* 29(1): 7–40.

1986 Koncepcja 'terytorium eksploatowanego przez osadę' w archeologii brytyjskiej i jej implikacje badawcze. *Archeologia Polski* 31: 7–30.

1987 Podstawowe metody analizy punktowych układów przestrzennych. *Archeologia Polski* 32: 21–53.

1997 Aktualne zagrożenia dziedzictwa archeologicznego w Polsce. In A. Prinke (ed.), *Aktualne zagrożenia dziedzictwa archeologicznego*, 7–25. Poznań.

Kobyliński, Z., and P. Urbańczyk
1984 Modelowanie symulacyjne pradziejowych procesów osadniczych. *Kwartalnik Historii Kultury Materialnej* 32: 67–94.

Kobyliński, Z., J. Zagrodzki and W. Brzeziński
1984 Apis: komputerowy program próbkowania losowego i statystycznej analizy przestrzennej stanowiska archeologicznego. *Archeologia Polski* 29: 41–55.

Koj, J.
1996 Ocena pierwszych lat badań AZP w województwie częstochowskim. In D. Jaskanis (ed.), 1996b: 103–107.

Konopka, M.
1981a Problem wdrożenia programu 'Zdjęcia archeologicznego' w Polsce — koncepcja realizacji. In M. Konopka (ed.), 1981c: 28–39.

1981b Instrukcja wypełniania karty ewidencji stanowiska archeologicznego. In M. Konopka (ed.), 1981c: 40–48.

1984a *Instrukcja ewidencji stanowisk archeologicznych metodą badań powierzchniowych (Archeologiczne Zdjęcie Polski).* Warszawa.

1984b Carte des sites archéologiques en Pologne, methodes et organisation. *Archaeologia Polona* 21/22 (1983): 187–216.

Konopka, M. (ed.)
1981c *Zdjęcie Archeologiczne Polski.* Biblioteka Muzealnictwa i Ochrony Zabytków seria B vol. LXVI. Warszawa.

Kostrzewski, J.
1950 W sprawie inwentaryzacji zabytków przedhistorycznych. *Z Otchłani Wieków* 19: 18–19.

Krawczyk, M., and W. Rominski
1994 Wyniki badań powierzchniowych na obszarze 93–32. In E. Tomczak (ed.), 1994: 127–37.

Kruk, J.
1970 Z zagadnień metodyki badań poszukiwawczych. *Sprawozdania Archeologiczne* 22: 445–56.

1973 *Studia osadnicze nad neolitem wyżyn lessowych.* Wrocław.

Kurnatowska, Z., and S. Kurnatowski
1996 Uwagi o AZP z perspektywy badań w wielkopolsce. In D. Jaskanis (ed.) 1996b: 79–91.

Larsen, C.U. (ed.)
1992 *Sites and Monuments Records.* Copenhagen.

Lloyd, J.A., and G. Barker
1981 Rural settlement in Roman Molise. Problems of archaeological survey. In G. Barker and R. Hodges (eds.), *Archaeology and Italian Society*, 375–416. British Archaeological Reports, International Series 102.

Łosiński, W.
1982 *Osadnictwo plemienne Pomorza (VI–X wiek).* Wrocław.

Matoga, A.
1996 Archeologiczne Zdjęcie Polski — polowa drogi. In Jaskanis 1996c: 47–61.

Mazurowski, R.
1980 *Metodyka archeologicznych badań powierzchniowych.* Warszawa-Poznań.

Moore, J.A., and A.S. Keene (eds.)
1983 *Archaeological Hammers and Theories.* New York: Academic Press.

Parczewski, M.
1991 *Początki kształtowania się polsko-ruskiej rubieży etnicznej w Karpatach.* Kraków.

Prinke, A.
1996 *AZP-Fox, Release 1.8: A Computer Database Management System on Archaeological Sites. User's Guide.* Poznań.

Rózycki, A.
1980 Pojęcie zjawiska koncentracji w polu rozrzutu zabytków ruchomych na stanowiskach archeologicznych. *Archeologia Polski* 24: 21–52.

Rydzewski, J.
1986 Przemiany stref zasiedlenia na wyżynach lessowych zachodniej Małopolski w epoce brązu i żelaza. *Archeologia Polski* 31: 125–94.

1996 Archeologiczne Zdjęcie Polski — doświadczenia i perspektywy. In D. Jaskanis (ed.) 1996c: 62–78.

Tomczak, E. (ed.)
1994 *Badania archeologiczne na Górnym Śląsku i w Zagłębiu Dąbrowskim w latach 1991–1992.* Katowice.

Wiślański, T.
1969 *Podstawy gospodarcze plemion neolitycznych w Polsce północno-zachodniej.* Wrocław.

Woyda, S.
1974 O pracach nad zdjęciem archeologicznym terenu Mazowsza i Podlasia. *Wiadomości Archeologiczne* 39(1): 44–47.

1981 Archeologiczne zdjęcie terenu — ogólne założenia metody w oparciu o doświadczenia mazowieckie. In H. Konopka (ed.) 1981c: 11–21.

Zin, W.
1980 *Principles of the Realisation of the Archaeological Field Record of Poland.* Document issued on 15 February 1980 by Professor dr Wiktor Zin, Vice Minister of Culture and Arts, and Conservator-General.

8. Surveying Prehistoric Industrial Activities: The Case of Iron Production

Evžen Neustupný and Natalie Venclová

Summary

Special attention should be paid to the relics of prehistoric industrial activities during fieldwalking. Manufacturing waste, which represents the most frequent type of finds connected with production activities, behaves differently from pottery sherds in the ploughsoil. Ecofacts with 'absolute frequencies' that do not disintegrate, such as iron bloomery slag or fragments of sapropelite, prove to be most helpful in studying ancient production (Neustupný and Venclová 1996; 1998). Spatial distribution of production relics not only reflects patterns of residential activities, but also the 'industrial pattern' within a landscape. Surface prospecting aimed at recognizing specific production activities (e.g. iron production) has produced good results in the Loděnice project in west-central Bohemia. It promises even more when applied on a larger scale. Types of production activities could be safely identified by the manufacturing waste, but problems arose in dating it, especially in the case of iron slag. A most helpful tool in this field is offered by statistics and multivariate methods combined with GIS procedures (cf. Neustupný 1996).

Introduction

Prehistoric production activities (or, more exactly, remains of such activities) became the subject of archaeological concern long ago. The same, however, cannot be said about the wider spatial context of the remains. Results from studies of individual types of artefacts, workshops, production technology and techniques of manufacture are frequently available, but information is missing on the relationships of production and residential areas, on the spatial distribution of workshops and on the organization of production in general. In short, production has mostly been studied in isolation, and only rarely together with the corresponding settlement patterns of larger regions. As a result, the integration of production activities into their spatial context, or the industrial pattern of a region, have hardly been touched.

The production of iron was chosen in the present case-study to demonstrate the potential of surface artefact survey for the study of industrial aspects of human settlement, or of the prehistoric industrial landscape.

Successful surveys of prehistoric or medieval iron-producing regions have been carried out in various parts of Europe (e.g. Pieta 1989 with refs., or Roth 1995 for Slovakia; Gömöri 1995 for Hungary; Jockenhövel and Willms 1993 for Germany, to cite just a few examples from Central Europe), but the survey was, up to now, carried out mostly by traditional methods aimed at finding 'sites' and not paying attention to their spatial and settlement relations within the study regions. The latter goal was addressed in a fieldwalking project described below.

A short review is necessary of the record of prehistoric iron manufacture in Bohemia and especially in the study region. Evidence of iron production is available for the Hallstatt (isolated finds only), La Tène (6–8 sites in Bohemia with finds of furnaces, and a few bloomery slag finds) and Roman periods (a larger number of small workshops within rural settlements). For the La Tène period, the small number of production sites may seem somewhat puzzling if contrasted with the large quantity and rich assortment of iron weapons, tools and ornaments found in graves of that period. The absence of evidence for a trade in iron (ingots, bars, etc.) makes large-scale importation of iron improbable. The slag pit furnace has been assumed to represent a typical La Tène iron-smelting furnace (Pleiner 1993; 1994). On very rare occasions though, the shallow open-hearth type also came to light (Venclová et al. 1998; Pleiner 1998). It should be mentioned that both of the finds of the latter type known up to now came to light in the study region.

A new approach to the archaeological record and its analysis (Venclová 1995; 1999) is apparently needed. It is based on the assumption that the small number of La Tène bloomeries may not be surprising, if the following facts are considered:

1) Not only sunken, but also (or mainly?) shallow hearth-type furnaces and other shallow production features can be assumed for the La Tène period.

2) Such features, being less preserved, may be totally destroyed by ploughing, or they may remain unobserved.
3) As much of the work connected with iron smelting must have been performed on the ancient surface, the evidence for it can be expected near to the surface, and mainly in the ploughsoil.

These considerations, among others, have initiated the surface artefact survey in the Loděnice region.

The survey of industrial relics is strongly biased in one respect. While the recording of a production area may be relatively easy when an appropriate field-walking method is applied — and in the case of iron smelting it is facilitated by the ease with which bloomery slag is identified — any direct dating of the production waste is impossible. This handicap may, however, be overcome to a certain extent by statistical means and multivariate handling of spatial relations of surface finds followed by GIS representation, as described below.

Slag and its behaviour in archaeological contexts

It seems appropriate to begin with a few general comments on the properties of bloomery slag and its behaviour in archaeological contexts. Slag belongs to

'absolute' ecofacts, i.e. archaeological relics that do not disappear easily from the archaeological record, and, not being subject to the reduction transformation on any large scale, they are not reduced in quantity (cf. Neustupný 1993: 55; Neustupný and Venclová 1996: 616–17). Typical bloomery slag, the product of the direct reduction of iron (smithing slag or any atypical slags are not considered in the following analysis) is highly visible on the ploughsoil surface. The mobility of the rather heavy slag lumps in the ploughsoil is not high, and the pieces are not transported far away during post-depositional processes. According to the data presently available in Bohemia, the accumulation (heaping) of slag does not make it conspicuous in the field. Prehistoric slag was most probably not recycled in Post-Mediaeval iron works as was the case of large prominent slag heaps reported for the Mediaeval (?) period, for example, in the Krušné Hory mountains in northwest Bohemia (Kořan 1969: 5–6; Maur 1995), the Forest of Dean in Britain (Bick 1990) or Schwäbische Alb in Germany (Kempa and Yalçin 1994), and so on.

The following comments on the behaviour of pieces of slag on the ploughsoil surface, in the ploughsoil and in the subsurface features are based on the results of different types of field activities (including excavations), but above all on the record obtained by the Loděnice fieldwalking project (Figure 8.1) described in detail below.

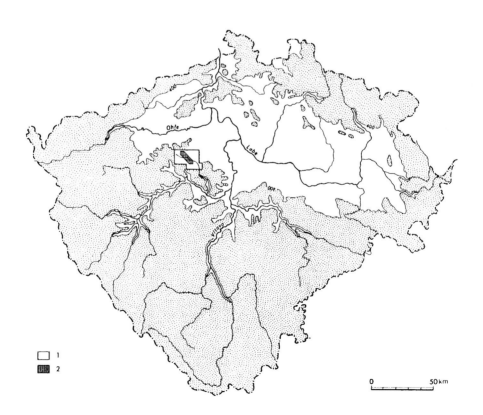

Figure 8.1   The Loděnice Project region in Bohemia. Hatched area is that of the analytical surface survey.

Slag on the ploughsoil surface

It was observed during fieldwalking that finds of slag formed more or less continuous areas on the ploughsoil surface, extending up to 3 ha, rarely 4–5 ha or more. The largest area continuously covered by slag consisted of 17 fieldwalking squares (hectares, in this particular case). Where pieces of slag were scattered more widely, their density was lower, sometimes being represented by single items. Even under the best fieldwalking conditions, the quantity of slag did not exceed 397 pcs/ha in the heaviest accumulations (but it should be kept in mind that concentrations of this intensity originated from only 10% of the surface surveyed). In no case was a continuous compact layer, let alone a heap protruding above the present surface, observed. This may be conditioned by the fact that the survey was carried out exclusively on cultivated soil, where the original accumulations were repeatedly disturbed by ploughing and other farming activities. Nevertheless the evidence seems to be convincing that very large slag heaps did not exist in the study region, in contrast to those characterizing the large-scale production in the Roman period bloomery districts of the *barbaricum*, in the Classical world or in the High Middle Ages (cf. e.g., Sablayrolles 1982; Jockenhövel and Willms 1993; Dumasy 1994; Voss 1988 for slag heaps and their secondary use). There are at least two possible explanations for this assumption: (1) Ploughing would probably avoid very large slag accumulations, which would therefore survive up to now or, alternatively, they would be ploughed away leaving

concentrations of slag much richer than those actually recorded. (2) Heaps of ancient slag with high iron content would most probably not escape the attention of smelters of the historical period. As a rule, they would be recycled, that is, used as a source of iron, and this would not go unobserved in written records.

In spite of the absence of large accumulations of the above-mentioned type, prehistoric iron slag represents a highly visible object, easily identifiable by any fieldwalker after a brief training.

The location of iron production workshops is reliably indicated by high accumulations of slag. It is a consequence of the characteristic properties of slag, which can be classified as primary refuse, found on the spot (or in the production area) where it was produced (cf. Schiffer 1987: 58–59). This has been confirmed in the study region by archaeological investigations combined with magnetometric prospection and magnetic susceptibility measurements, reflecting, though, not only slag but other magnetic materials as well (cf. Fröhlich *et al.* 1998). The bloomery workshops excavated up to now have all been reflected in marked cumulations of slag on the ploughsoil surface. Examples are presented in Figures 8.2 and 8.3.

Slag in the ploughsoil

While it has been presumed that *c.* 0.3% to 16.6% of pottery, lithics, etc. out of their total amount in the ploughsoil may be found on the ploughsoil surface (Kuna 1994: 33–34 with refs.; Vencl 1995: 24–26 with

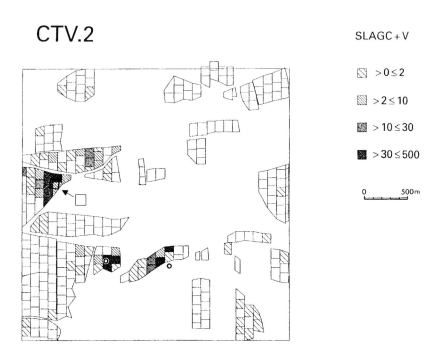

Figure 8.2   The Loděnice Project: the quantity of bloomery slag pieces on the ploughsoil surface in the polygons surveyed compared to the location of iron-smelting furnaces excavated (circles) or presumed (square, cf. Figure 8.3).

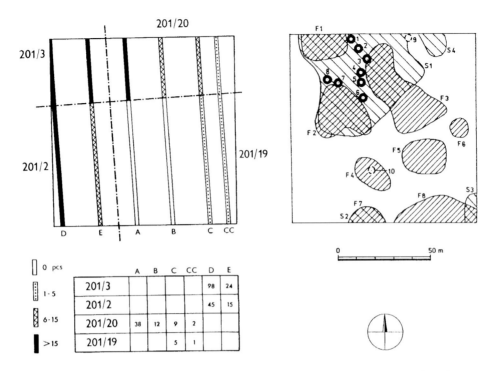

		A	B	C	CC	D	E
	201/3					98	24
	201/2					45	15
	201/20	38	12	9	2		
	201/19			5	1		

Figure 8.3   The La Tène period industrial site Mšec I: the fieldwalking results (left) compared to the magnetometric prospection, magnetic susceptibility measurements and soil phosphate analysis (right; after Fröhlich *et al.* 1998). The quantity of slag pieces in the lines fieldwalked is indicated. 1–10 geomagnetic anomalies, S1–S4 areas of high magnetic susceptibility, F1–F8 high soil phosphate contents.

refs.), it seems likely that this percentage is higher in the case of bloomery slag. This observation is based on the average dimensions of slag lumps, usually exceeding those of the other types of finds. Nevertheless, precise data are lacking.

The relation of the ploughsoil contents to subsurface features has been studied in the course of the excavations of the Iron Age (La Tène period) industrial settlement at Mšecké Žehrovice I (Venclová *et al.* 1998). The two highest accumulations of bloomery slag in the ploughsoil, that is, over 6 kg per cubic meter of earth (up to more than 10 kg/cu m) corresponded to shallow production features and their close surroundings (one example: Figure 8.4, A). In other words, the correlation of the above-average quantity of slag in the ploughsoil to the subsurface production features was striking. Some other production features, however, were not reflected by surface scatters. The assumed reason was as follows: either there were layers of soil between the feature fills and the ploughsoil (Figure 8.4, B and E), or the dimensions of the production feature and its slag content were small, so that the slag concerned did not exceed the background values of slag contents in the ploughsoil. (Figure 8.4, D). (In the area of the described industrial site very low 'background' values of slag were recorded in almost all archaeological trenches.) It should be stated that production installations excavated at the site in all cases represented shallow

or non-sunken features; they did not consist of deep pits, typical, for example, for slag pit furnaces.

It can be assumed that high slag densities in the ploughsoil (and on its surface) originate, on the one hand, from ploughed-away shallow sunken features filled with slag (open-hearth furnaces or reheating hearths), and on the other hand from slag heaps (waste accumulations). The low slag densities in the ploughsoil need not come from such features, but they may possibly originate from near-by activity areas located on the ancient surface and exposed to later ploughing (Figure 8.4, C). As is well known, a part of the activities connected with the production of iron was carried out outdoors, that is, on the ancient surface.

Slag in subsurface features

It seems logical to expect slag in production features, that is, in pits of the slag pit furnaces, in reheating hearths and in the deposits of waste; some of the latter could have been placed in some sunken features. Slag, however, need not be found in slag pit furnaces at all: it is well known that furnaces with their slag content preserved in slag pits are found less frequently than empty ones. In the La Tène period bloomery plant in Mšec III, slag blocks were found in only 2 furnaces out of 18 (Pleiner and Princ 1984); in the Roman period workshop in Praha-Dubeč, none

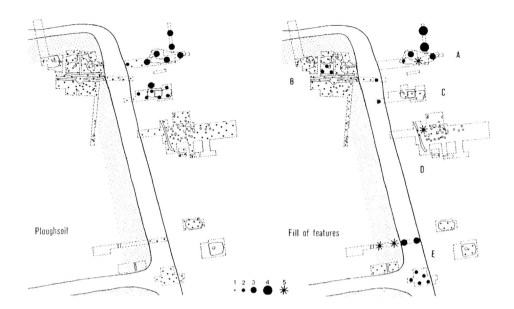

Figure 8.4   The La Tène period industrial site Mšecké Žehrovice I: bloomery slag in the ploughsoil and in the fill of features. The later enclosure by banks and ditches marked by dots and full lines. Dotted lines, archaeological trenches. A, industrial feature with a cumulation of slag; B, slag block under the settlement layer; C, settlement feature with no industrial installation; D, small reheating hearth; E, accumulation of slag in the bank and ditch of the enclosure built later on the industrial site. Density of slag in kg/cub m: 1 >1 kg, 2 ≥1–1.99 kg, 3 ≥2–5.99 kg, 4 ≥6–9.99 kg, 5 ≥10 kg.

was left in the 7–8 furnaces excavated (Vencl *et al.* 1976); 7 slag blocks were gained from 19 excavated furnaces in Říčany (Kuna *et al.* 1989); and the site of Ořech produced 11 blocks out of 44 furnaces (Motyková and Pleiner 1987). This means that a large proportion of slag was removed from the furnaces (by the smelters themselves, if the furnace was to be used repeatedly) and deposited somewhere in the vicinity — in some of the abandoned working pits or on the ancient surface. In cultivated areas, this slag can hardly be found outside modern ploughsoil.

According to the excavations at Mšecké Žehrovice site I the ploughsoil in areas located on ploughed fields contained 32.5% (132.5 kg) of the total volume of the slag found, 61.2% (249.5 kg) coming from the fill of production features and only 6.3% (26.0 kg) from the fill of other (non-production) features.

The Loděnice Project

The fieldwalking project in west-central Bohemia (referred to, hitherto, as the 'Loděnice Project') was carried out in 1991–95. The study region covered a territory of 253 sq km. The whole area was surveyed by a variety of traditional methods, and 45 sq km of it were submitted to the analytical type of survey, that is, such that subdivides the whole area into parts surveyed separately (cf. Neustupný 1998 for the term). The region is a watershed plateau surrounded by basins of the uppermost courses of three small streams: Loděnice in the south, Bakov(ský) in the north and Červený in the east, only the first two mentioned being discussed in this paper. The two main stream valleys show contrasting natural conditions. The northern, narrow and steep-sloped valley represents the area of outcrops of the Kounov sapropelite (black material used for the manufacture of rings in the La Tène period), while the southern part is constituted by a broad valley with mild slopes and no sapropelite sources. In both parts of the region, a surface occurrence of pelosideritic concretions has been identified, which are presumed to represent the raw material for the prehistoric production of iron.

The area is believed to represent a western extension of the richly settled region of central Bohemia, as the areas to the immediate north and west seem to have been uninhabited in prehistoric times. This fact is usually attributed to the fairly high elevation above sea level (more than 400 m in the case of the Loděnice basin). The area to the immediate east (the Kladno district) was one of the industrial centres of Bohemia until quite recently (coal mining and iron production based on local resources).

Excavations executed in the region (Pleiner and Princ 1984; Venclová *et al.* 1998; Zeman *et al.* 1998) have demonstrated production of iron in the La Tène and Roman periods. Being limited in spatial cover, however, these results could not easily be transferred to the whole region. The obvious requirement was to look at the region as a whole.

The field project was carried out by one of the present authors in 1993–95 (Venclová 1995 with refs.). The method adopted was basically that proposed by M. Kuna for the ALRNB project (Kuna *et al.* 1993; this volume). It consisted of surveying in randomly selected polygons, each approximating to 1 sq ha. The distance between the lines walked was 20 m. All artefacts and ecofacts, prehistoric to post-medieval, were collected. These basic polygons have been assembled into larger polygons without leaving empty zones between them; this tactic proved to have many advantages. Only deforested arable land was surveyed.

The method has been described separately (Venclová *et al.*, in press). The evidence coming from approximately 1350 1 ha polygons was supplemented by the results of earlier excavations and by museum finds; the assembled evidence has been made available in the form of databases that could serve directly as input into software like IDRISI and SPSS.

The fieldwork was followed by detailed description and classification of the finds, deposition of the resulting information in databases and by a rather complicated mathematical (partly statistical) computerized research, using methods of which some were specially developed for this project (Neustupný and Venclová 1996). The main constituents of this elaborate methodology were vector synthesis (in the form of principal component analysis; cf. Neustupný 1993) and Geographical Information Systems (or GIS; the raster IDRISI software was used). What we consider to be crucial in our method is the combination of multivariate procedures with the GIS software, as described elsewhere (Neustupný 1996). One may note that this seems to be an ideal method for the evaluation of any fieldwalking results and it has already been applied successfully to projects other than Loděnice (Kuna 1997; this volume).

The evidence

Table 8.1 demonstrates the frequencies of the main categories of finds of (possibly) the La Tène period

found in the project. The mass occurrence of pieces of slag is obvious. But what age are they? As stated above, iron production in the region is also proved for at least the Roman Period by archaeological excavations, and it cannot be excluded that iron was also produced there in other periods.

Chronology

The pieces of slag obtained by fieldwalking could not be dated contextually, and there is no typology of this kind of ecofact that would allow their chronological classification. As the slag came to light in 1 ha basic polygons, it became possible to formulate spatial relationships between the slag and other categories of finds. There are four ways to study these relationships.

(1) We have calculated the usual (product-moment) correlation coefficient between slag and the other major categories of finds (cf. Table 8.2) based on their common occurrence in the basic polygons. The number of fragments was used in the calculations. It was found that slag had highly significant correlations with La Tène pottery, with cores originating in the production of sapropelite bracelets (SAPCO), and with worked sapropelite in general (SAPW; these two last named categories date to the La Tène period because sapropelite rings are found exclusively in graves of one of its subperiods). Also highly significant was its correlation with prehistoric pottery in general (i.e. pottery not datable more exactly), which is a natural consequence of the categorization. The correlation with post-mediaeval finds was high, but the coefficient with Roman period pottery was not statistically significant at the 5% level, though a slight increase of the level would make it significant.

(2) The second kind of evidence that can throw some light on the chronology of slag finds is the distance from polygons containing pieces of slag to polygons containing La Tène pottery, Roman period pottery and sapropelite (Table 8.3). Not only is the

Descriptor	Pieces	Maximum	Average	STD	Polygons
Pottery	183	10	0.13	0.62	101
SAPCO	318	23	0.23	1.33	112
SAPW	1152	40	0.85	2.58	357
SAPN	1438	27	1.06	2.42	477
SAPV	8892	580	6.53	40.05	89
S_SAP	11800	605	8.67	42.37	593
S_SLAG	3358	397	2.47	20.03	282

Table 8.1   Frequencies of descriptors of the La Tène period. SAPCO = sapropelite cores; SAPW = worked sapropelite; SAPN = unworked sapropelite; SAPV = visible but uncollected pieces of sapropelite; S_SAP = all sapropelite categories; S_SLAG = all slag categories.

Find category	Correlation with SLAG	Level of significance (P =)
SAPCO	.0958	.000
SAPW	.1807	.000
SAPN	.0625	.021
SAPV	.0102	.706
Neo-Eneolithic	−.0048	.860
Bronze Age	−.0062	.819
Hallstatt period	−.0020	.941
La Tène period	.2438	.000
Roman period	.0516	.057
Farming prehistory	.2800	.000
Chipped industry	−.0096	.722
Early Medieval	−.0029	.914
High Medieval	.0103	.704
Post-Medieval	.0675	.013

Table 8.2   Correlation coefficients and significance levels between slag and some other find categories. For key see Table 8.1.

average distance from Roman pottery far greater than from the La Tène period finds, but this difference increases with the number of slag fragments found. The information on distances can easily be obtained by means of GIS software.

Table 8.3 demonstrates that (1) the polygons containing slag are nearest to the polygons containing sapropelite, and this is valid irrespective of the frequency of slag fragments; (2) the polygons containing La Tène pottery come next; (3) the average distance of Roman period polygons from polygons with slag is much greater, and this is again valid irrespective of frequency; (4) polygons containing less slag are substantially more distant from those with pottery and sapropelite than polygons rich in slag; (5) the progression described in paragraph 4 below is much stronger with sapropelite and La Tène pottery than with Roman pottery.

(3) We have published a map (Neustupný and Venclová 1996: 718) showing how different the spatial distribution of slag is from the distribution of Roman period finds. Another form of the map is published here (Figure 8.5). However, such maps, in spite of being nearer to traditional archaeological reasoning,

may sometimes be misleading, because a graphical representation of this kind does not take into consideration important quantitative factors. Be that as it may, it is obvious from the map that there are important concentrations of slag not accompanied by any Roman period finds and vice versa.

(4) The fourth kind of evidence, and a very strong one, is the clustering of slag (Neustupný and Venclová 1996: 717) exactly in those areas where La Tène pottery, SAPCO and SAPW also cluster (Neustupný and Venclová 1996: 724). The correspondence is not complete, but in no case could the differences be explained by redating the slag to the Roman period (Figure 8.5).

All this evidence does not entirely exclude the possibility that some pieces of slag were in fact produced in the Roman period or even later, but it certainly favours the dating of most pieces of slag to the La Tène period.

If shallow open-hearth-type furnaces prevailed in the La Tène period (in contrast to the deeper slag pit furnaces of the Roman period), it would seem logical that widespread scatters of slag fragments in the ploughsoil was more typical of La Tène. Of course, the dispersion also depends on factors other than the type of furnaces, and the whole issue should be studied on empirical grounds.

Regionalization

The Loděnice region is rather small: the area shown in the maps is about 65 sq km, but only some 43 sq km were included in the project, which covered 11.5 sq km by fieldwalking. Nevertheless the area of the project consisted of two basins separated by a sharp divide. The northern part, the basin of the Bakov stream, can be characterized by deep narrow valleys with unusually steep slopes, while the southern area, that of the Loděnice stream itself, was much more even. The Bakov basin was the place where the outcrops of sapropelite could be exploited (Břeň 1955; Venclová 1992; 1994; Venclová *et al.* 1998).

We studied the whole area of the project by means of vector synthesis, using the principal component algorithm with Varimax rotation of three factors (for

No. of pieces of slag in polygon	La Tène pottery	Sapropelite	Roman period pottery	No. of polygons
0	1255.00	871.00	1807.00	1079
1–10	249.79	42.97	887.65	242
11–27	110.14	45.00	603.95	22
40–397	45.00	2.56	407.39	18

Table 8.3   Average mean distance from polygons containing slag to polygons with other kinds of finds (in metres).

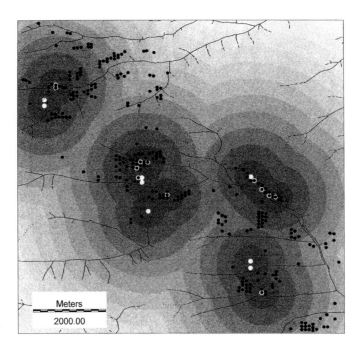

Figure 8.5   Loděnice region: white circles, polygons with Roman period finds; shaded zones, distances from such finds (band width approx. 500 m); black dots, polygons with finds of iron slag.

the method cf. Neustupný 1996; 1997; Neustupný and Venclová 1996). The results are displayed in Table 8.4.

Factor 2 is mainly characterized by unworked sapropelite (SAPN) and worked sapropelite (SAPW), whilst the cores of sapropelite rings (SAPCO) and all slag categories (S_SLAG) are much less relevant, and pottery (LATEN) is quite irrelevant. High factor scores in respect of this factor cluster in the Bakov stream subarea (Neustupný and Venclová 1996: 720). Clearly, this factor has to do with the exploitation and the primary shaping of sapropelite.

The loadings of Factor 1 are high in the case of SAPCO, SAPW and LATEN pottery. The scores of this factor (Neustupný and Venclová 1996: 719) clearly point to residential areas with domestic production of sapropelite rings, both in the basin of the Loděnice stream and in the Bakov stream basin.

Factor 3 is most significantly connected with iron slag, but La Tène pottery has a strong loading in

respect of this factor as well. In view of the strength of correlations in the original matrix, the loading for SAPW may not be entirely due to chance. Factor scores in respect of Factor 3 when displayed by means of GIS (Neustupný and Venclová 1996: 721), show high values (exceeding 1) exclusively in the Loděnice basin. Of course, iron was also produced in the Bakov basin, but apparently with lower relative intensity.

If the number of slag pieces (SUM), their total weight (WEIGHT) in individual polygons and their average weight (AV_WEI = WEIGHT/SUM for each polygon) are considered separately for the two basins (cf. Table 8.5), we get diverging values. Both the average number of pieces and their weight in the polygons are much less in the case of the Bakov stream than in the Loděnice basin, and these differences are statistically highly significant (the level of significance is 0.02 for SUM and 0.011 for WEIGHT, variances being homogeneous). In the case of average weight (AV_WEI), however, the variances are not homogeneous, and the difference is not statistically significant below the level of 20%. These facts strengthen the results of vector synthesis as described above, supporting them by a statistical argument.

Table 8.6 displays a selection of environmental variables contrasted for the two watersheds. The differences for the two basins are highly significant in all the four cases, but the variances are not homogeneous in the case of slope (SLOPE) and distance from the nearest running water (D_VODA), which makes the results somewhat shaky. If a *t*-test is used, however, the results remain highly significant even if

Category	Factor 1	Factor 2	Factor 3
LATEN	.68737	−.07438	.38479
SAPCO	.85588	.06340	−.09082
SAPW	.69719	.51518	.10173
SAPN	.04151	.95904	.03106
S_SLAG	.06551	.06173	.95589

Table 8.4   Factor loadings of the first three principal component factors (Varimax rotated).

		Watershed	*N*	Mean	Standard deviation
SUM	WSH	1	187	4.66	8.52
		2	94	2.53	2.93
	Total		281	3.95	7.21
WEIGHT	WSH	1	187	226.96	368.28
		2	94	118.00	267.36
	Total		281	190.51	341.31
AV_WEI	WSH	1	187	61.602	69.557
		2	94	45.572	140.481
	Total		281	56.240	99.127

Table 8.5   Average number and weight of slag pieces in polygons in watershed 1 (Loděnice) and watershed 2 (Bakov). The same for the average pieces of slag in each polygon.

		Watershed	*N*	Mean	Standard Deviation
SLOPE	WSH	1	187	3.00	1.99
		2	94	5.78	2.09
	Total		281	3.93	2.41
D_SLOPE6	WSH	1	187	253.20	206.48
		2	94	32.72	45.34
	Total		281	179.44	199.66
D_VODA	WSH	1	187	221.33	131.60
		2	94	162.80	114.59
	Total		281	201.75	128.96
VYSKA	WSH	1	187	417.66	12.90
		2	94	369.62	18.67
	Total		281	401.59	27.24

Table 8.6   Environmental variables for polygons containing slag (S_SLAG). Watershed 1 (Loděnice) and 2 (Bakov) are contrasted. D.SLOPE6: distance from slope of 6° or steeper, D_VODA: distance from the nearest stream, VYSKA: altitude, WSH: watershed.

equal variances are not assumed. Similar results are obtained for variables other than S_SLAG. Therefore, the differences set out in Table 8.6 are conditioned by the generally differing natural conditions in the two basins rather than by any special position of slag among the variables.

This is a manifestation of regional variability in the two basins, in itself a prolongation of the situation that can also be detected in earlier phases of prehistory (cf. Neustupný 1998).

Specialization of communities

The results of vector synthesis discussed above also demonstrate the inherent diversity within the Loděnice area as a whole. This is manifested by Factor 1 and Factor 2 that both display high loadings towards pottery (LATEN). In the former case, however, pottery is accompanied by worked sapropelite (SAPCO and SAPW), while in the latter case it is associated with slag. As pottery is invariably a marker of residential areas, this seems to mean that there were two kinds of residential areas and, consequently, two specializations for Iron Age communities: one characterized by the production of iron, the other by the manufacture of sapropelite rings.

Displaying the factor maps of Factor 1 and 3 synoptically (Figures 8.6 and 8.7) we can easily distinguish the two types of residential areas, as well as the third type in which finds of sapropelite and of slag are balanced.

Statistical tests that could bear on the statistical significance of the differences have not yet been performed, but at least some of the differences seem

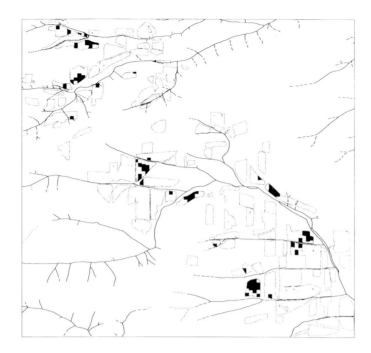

Figure 8.6   Loděnice region: factor scores of Factor 1 (residential areas).

Figure 8.7   Loděnice region: factor scores of Factor 3 (production of iron).

to be valid. In spite of the fact that the contrasts need not be absolute, they are extremely important. This is a serious indication that individual Iron Age villages lying at a distance of a few kilometres from each other could have had different economies.

This result is only slightly influenced by the possible objection that the two economic orientations need not be strictly contemporaneous, in view of the inherent difficulties of dating archaeological artifacts. Even if they were not used at the same time — something that we do not consider to be very likely — it would mean that they followed each other within a short timespan. Thus, they would testify to the dynamics of the orientation of individual communities. It does seem that we are reaching the genuine individuality of Iron Age communities that has some real counterpart in the archaeological record.

Conclusions

The brief comments on the behaviour of some industrial relics in different archaeological contexts, and the results of surface survey within the Loděnice Project, have demonstrated not only the suitability, but rather the indispensability of surface survey for the study of prehistoric industrial activities.

The number of iron production components recorded in the study region markedly exceeds those known from subsurface finds. This is logical because a large part of the evidence for bloomery workshops (as well as other production areas) is only preserved near to the present surface: most of the production activities were carried out on the ancient surface, and most of the waste has been left there. In the conditions of the intensively cultivated Bohemian landscape — the same being true of other landscapes in Central Europe — it is the ploughsoil that offers the bulk of the evidence for artefacts and ecofacts connected with iron making.

The analytical type of surface survey is of primary importance for the study of production and its relation to settlement. Whilst traditional fieldwalking was able to find just the very conspicuous scatters of industrial finds, or 'sites', analytical survey methods also record and quantify the less clustered relics, and the information supplied by them is much more complex.

The Loděnice Project has brought a wealth of information on early iron metallurgy. We could suggest that there were two subregions in the project area that differed in relation to production activities, and that not even all communities in the same subregion played the same role in Iron Age social networks. Thus, we consider it possible to come to the individuality of prehistoric events on a basis different from empathy. This opens the way to sound theorizing rather than to endless repetition of rationalistic assumptions.

It is difficult to hypothesize how much it would be necessary to excavate if the same results should be achieved by means of the traditional 'destructive' strategy. It was easy and very fast to achieve all this on the basis of a well-targeted and theoretically well-founded fieldwalking project.

We believe, however, that sound theory preceding field research is not enough. We can hardly imagine the results without rather complicated methodology beginning with the 'analytical' method of fieldwalking (cf. Kuna 1998). This method generates so much information (so many units of observation) that traditional mapping of empirical observations may not represent relevant procedures for grasping the structures contained in the record.

We are firmly convinced that all the information can hardly be extracted from the finds without non-trivial manipulation with databases, that is, without the use of rather complicated statistical tools and the application of vector synthesis ('multivariate analysis'), as well as without Geographical Information Systems. Moreover, all these procedures must be closely integrated.

This brings archaeology to a new state that is both theoretically and methodologically much more demanding than the traditional narrations. It has no sense to pretend that archaeology can be at one and the same time a scientific discipline and a pastime for the general public. Archaeology becomes an occupation for educated professionals that cannot be equated with the products by means of which it addresses its discoveries to non-archaeologists.

Acknowledgements

This paper presents some of the results of the project supported by the Grant Agency of the Czech Republic (registration no. 404/97/K024).

References

Bick, D.
  1990   Early iron production from the Forest of Dean and district. *Historical Metallurgy* 24(1): 39–42.
Břeň, J.
  1955   Černé (švartnové) náramky v českém laténu (Fabrication de bracelets en sapropélite en Bohême). *Sborník Národního musea v Praze IX-A Historia* no. 1: 3–42.
Dumasy, F.
  1994   La métallurgie du fer dans la cité des Bituriges Cubi. In M. Mangin, (ed.), 1994: 213–22.
Fröhlich, J., A. Majer and N. Venclová
  1998   Archeologická prospekce a průzkum měřením magnetické susceptibility zemin (Archaeological prospection and measurements of magnetic susceptibility of soils). In P. Kouril, R. Nekuda and J. Unger (eds.), *Ve službách archeologie*, 87–93. Brno: Institute of Archaeology.
Gömöri, J.
  1995   Novšie archeologické výskumy nálezísk trosky z výroby železa v Maďarsku. *Študijné zvesti* 31: 231–41.
Jockenhövel, A., and Ch. Willms
  1993   Untersuchungen zur vorneuzeitlichen Eisengewinnung und Verarbeitung im Lahn-Dill-Gebiet. Ausgangslage und Ergebnisse der archäologischen Geländeprospektion. In H. Steuer and U. Zimmermann (eds.), *Montanarchäologie in Europa*, 517–29. Sigmaringen.
Kempa, M., and Ü. Yalçin
  1994   Exploitation des mines et métallurgie dans le Jura souabe. In M. Mangin (ed.), 1994: 227–35.
Kořan, J.
  1969   Vývoj železářství v Krušných horách. *Sborník Národního technického muzea v Praze* no. 8.

Kuna, M.
1994 *Archeologický průzkum povrchovými sběry* (Archaeological survey by surface collection). Zprávy České archeologické společnosti, Supplément 23.
1997 Geografický informační systém a výzkum pravěké sídelní struktury. In J. Macháček (ed.), *Počítačová podpora v archeologii*, 173–94. Brno: Ústav archeologie a muzeologie, Filozofická fakulta Masarykovy univerzity.

Kuna, M., J. Waldhauser and J. Zavřel
1989 *Říčany 1986. Záchranný archeologický výzkum sídliště doby laténské a železářského areálu starší doby římské* (Říčany 1986. Archäologische Rettungsgrabung einer latènezeitlichen Siedlung und eines Eisenverhüttungsareals der älteren römischen Kaiserzeit). Brandýs nad Labem — Stará Boleslav: Okresní muzeum Praha-východ.

Kuna, M., M. Zvelebil, P.J. Foster and D. Dreslerová
1993 Field survey and landscape archaeology research design: methodology of a regional field survey in Bohemia. *Památky archeologické* 84: 110–30.

Mangin, M. (ed.)
1994 *La sidérurgie ancienne de l'Est de la France dans son contexte européen*. Annales littéraires de l'Université de Besançon 536.

Maur, E.
1995 Přísečnické železářství v údobí přímé výroby železa. *Historický obzor* 11–12: 257–58.

Motyková, K., and R. Pleiner
1987 Die römerzeitliche Siedlung mit Eisenhütten in Ořech bei Prag. *Památky archeologické* 78: 271–448.

Neustupný, E.
1993 *Archaeological method*. Cambridge: Cambridge University Press.
1996 Polygons in archaeology. *Památky archeologické* 87: 112–36.
1997 Uvědomování minulosti (The cognizance of the past). *Archeologické rozhledy* 49: 217–30.

Neustupný, E. (ed.)
1998 *Space in prehistoric Bohemia*. Praha: Institute of Archaeology.

Neustupný, E., and N. Venclová
1996 Využití prostoru v laténu: region Loděnice (Gebrauch des Raumes in der Latènezeit: die Region Loděnice). *Archeologické rozhledy* 48: 615–724.
1998 The Loděnice region in prehistoric times. In Neustupný 1998: 81–102.

Pieta, K.
1989 Frühkaiserzeitliche Eisenverhüttungsanlage in Varín, Slowakei. In R. Pleiner (ed.), *Archaeometallurgy of iron 1967–1987*, 213–27. Prague: Institute of Archaeology.

Pleiner, R.
1993 The technology of iron making in the bloomery period. In R. Frankovich (ed.), *Archeologia delle attivitá estrattive e metallurgiche*, 533–60. Siena.
1994 Early bloomeries in Central Europe. In M. Mangin (ed.), 1994: 182–88.

1998 Production of iron at Mšecké Žehrovice. In Venclová, *et al.* 1998: 305–10.

Pleiner, R., and M. Princ
1984 Die latènezeitliche Eisenverhüttung und die Untersuchung einer Rennschmelze in Mšec, Böhmen. *Památky archeologické* 75: 133–80.

Roth, P.
1995 Metalurgie železa v dobe laténskej a rímskej na Spiši. *Študijné zvesti* 31: 105–22.

Sablayrolles, R.
1982 Intérêt et problèmes de l'étude des ferriers antiques: l'exemple de la Montagne Noire. In *Mines et fonderies antiques de la Gaule*, 183–90. Paris.

Schiffer, M.B.
1987 *Formation processes of the archaeological record*. Albuquerque: University of New Mexico Press.

Vencl, S.
1995 K otázce věrohodnosti svědectví povrchových průzkumů (Surface survey and the reliability of its results). *Archeologické rozhledy* 47: 11–57.

Vencl, S., N. Venclová and J. Zadák
1976 Osídlení z doby římské v Dubči a okolí (Eine römerzeitliche Besiedlung in Dubeč und Umgebung). *Archeologické rozhledy* 28: 247–78.

Venclová, N.
1992 Un atelier de travail du sapropélite à Mšecké Žehrovice en Bohême. In D. Vuaillat (ed.), *Le Berry et Limousin à l'Age du Fer. Artisanat du bois et des matières organiques. Actes du XIIIe Colloque de l'AFEAF*, 109–16. Guéret: Association pour la Recherche Archéologique en Limousin.
1994 The field survey of a prehistoric industrial region. *Památky archeologické, Supplementum* 1: 239–47.
1995 Specializovaná výroba: teorie a modely (Specialized production: theories and models). *Archeologické rozhledy* 47: 541–64.

Venclová, N.
1999 Iron production in the Loděnice region. *Nýchodoslovenský pravek*, Special Issue, 132–144.

Venclová, N. *et al.*
1998 *Mšecké Žehrovice in Bohemia. Archaeological background to a Celtic hero. 3rd–2nd cent. B.C.* Sceaux: Cronos Editions.

Venclová, N. *et al.*
in press *Sídlení a výroba. Projekt Loděnice* (Settlement and production: The Loděnice Project). Praha: Institute of Archaeology.

Voss, O.
1988 The iron production in Populonia. In G. Sperl (ed.), *The First Iron in the Mediterranean*. PACT 21: 91–100. Strasbourg.

Zeman, J., N. Venclová and J. Bubeník
1998 Železářská osada z 3.- poč. 5. stol. v Přerubenicích (Ein Siedlungs- und Eisenverhüttungsareal aus dem 3. Jh.- Anfang 5. Jh. in Přerubenice, Kr. Rakovník). *Praehistorica* 23: 95–131. Praha: Univerzita Karlova.

9. The Utility of the GIS Approach in the Collection, Management, Storage and Analysis of Surface Survey Data

Mark Gillings

Summary

Since the widespread adoption of GIS by archae-ologists in the early 1990s, GIS and regional survey have been seen as a highly complementary pairing. It is argued here that the ease with which GIS has been incorporated into the canon of intensive regional survey is due to a very particular theoretical con-ceptualization of what precisely GIS is. GIS as neutral tool. This has served to limit the impact of GIS approaches by restricting them to a set number of caricature roles: for example, the production of point-based distribution maps. GIS has much more to offer, but to realize its full potential, a clear recon-ceptualization has to take place on the part of survey practitioners. This can be stated quite simply: GIS as tool to GIS as approach. Such a reorientation has the potential to radically alter the way we approach existing problem areas, such as growing data complex-ity; normative rules of site definition; the incorpora-tion of geomorphological change; and the nature of heuristics, such as the manuring hypothesis. It can also create an ideal environment within which to develop new analytical approaches embracing developments in practice and theory, whether post-structuralist, Anna-liste or scientific.

In the light of this the aims of the present discus-sion are twofold. In the first instance to look at why survey and GIS have been so closely identified with one another and examine how this relationship has developed. In the second to suggest how the potential of an explicitly GIS-based approach may be most profitably exploited in the context of future survey theory and practice.

General introduction

Over the last three decades, the growing popularity of intensive regional survey has resulted in the amassing of a wealth of information ideally suited to the investigation of changing cultural landscapes. Land-scapes, it should be added, that are altering and developing over significant tracts of social space and time. Researchers have been fully aware that the data collected have the unique potential to enable broad spatio-temporal trends to be identified and explored, and, equally importantly, they can also permit areas at the immediate and local scale to be highlighted for more focused analysis and study. Despite this, at present the full range of possibilities and potentials for the detailed study of people–place relations encoded within these unique data sets is not being realized and exploited during the course of routine synthesis, analysis and interpretation. This, it will be argued, is a direct consequence of the complexity, diversity and problematic nature of much of this information.

One partial solution to this interpretive impasse, has been the advocacy of a set of computer-based techniques, grouped under the banner of Geographi-cal Information Systems, or, GIS. Notable within the literature of the last seven years has been a growing trend towards either the direct adoption of GIS or, more commonly, some acknowledgement of its per-ceived utility within the regional survey context.

The purpose of the present discussion is twofold. In the first instance the aim is to look at why survey and GIS have been so closely identified and how this relationship has developed, or, as some would argue, has signally failed to develop. Second, in developing the theme of the current volume, the goal will be to look ahead to how the potential of an explicitly GIS-based approach may be most profitably exploited in the future. The discussion will be structured into five broad sections. Following a general introduction, in the section on the survey context (p. 107) a number of fundamental points regarding the underlying nature of GIS and its relationship to regional survey will be raised and discussed. In the section on the current role of GIS (p. 108) the role allotted to GIS within regional survey projects will be analysed. In the final sections the evolving shape of future survey–GIS relationships will be explored and a number of gen-eral conclusions drawn.

The relationship between GIS and regional survey

It is widely accepted that the first major impact of GIS upon the wider archaeological consciousness came in 1990 with the publication of a collection of papers entitled *Interpreting Space* (Allen *et al.* 1990;

for a detailed historical discussion see Harris and Lock 1995: 350). Of the 12 case studies described in this ground-breaking volume, it is interesting to note that all had a regional landscape focus. In the introduction it was even asserted that *only* landscape-based archaeological research could provide the conceptual framework necessary to take full advantage of the potential the GIS methodology had to offer (Green 1990a: 5). As a result, it can be argued that it was at this point that the direct equation of GIS with regional landscape-based study became sedimented within current archaeological practice. This seminal volume was followed in 1991 by the publication of an equally influential book, focusing more explicitly upon the application of GIS within the established context of regional survey (Gaffney and Staňič 1991). As a result, GIS and surface survey, the latter contextualized within the more generic field of landscape archaeology, have been widely perceived as enjoying a special, if not exclusive, relationship. This optimistic view was shared by a number of researchers, with some even going so far as to suggest that the relationship was close enough to warrant marriage (Kvamme 1991: 12). Within this happy coupling, GIS was almost exclusively portrayed as a formidable panacea.

In retrospect it is easy to see why GIS and regional survey made such a visibly attractive combination. GIS applications, as portrayed in these ground-breaking case-studies, laid a clear emphasis upon spatial information, and engaged a map-like and highly visual approach to analysis and communication. They also employed a clear thematic structure based around the identification and conceptual isolation of discrete layers, and tended to be built upon the firm foundation of an environmental database. This has a particular immediacy for regional survey projects, whose theory and practice originated and developed amidst a highly dynamic theoretical climate. This was characterized by a renewed interest in quantitative methods and spatiality (e.g. Hodder and Orton 1976; Clarke 1977) and a strong emphasis upon environmental factors as integrated within a broader and more positivist systems-based paradigm (Alcock *et al.* 1994; Cherry *et al.* 1991: 14; Gibbon 1989: 76). More recently a number of particular readings of the highly diverse Annales historical approach (Bulliet 1992: 132; Last 1995: 143) have been endorsed and applied within a regional landscape context (Barker 1995; Bintliff 1991a; 1991b; Jameson *et al.* 1994). This perspective, with its advocacy of a 'total-historical' framework, stressing interdisciplinary collaboration and the application of the widest range of tools and approaches available, coupled with a Braudelian emphasis upon long-term, large-scale transformations, has proven equally complementary to the conceptual structure and capabilities of GIS (see Bintliff, this volume, for a discussion of the Annales model of history).

Given the impact within landscape studies, the fact that GIS applications have been a rare feature within individual site-based studies is striking. The dearth of applications of GIS at the intra-site scale, and repeated failure to cope adequately in the context of excavation, can be attributed to the failing of existing data models to accommodate the truly three-dimensional data generated by such enterprises (Harris and Lock 1996; cf. Green 1990b: 360).

The overall result of these parallel trends has been effectively to subsume GIS within a broad landscape remit, a position that has been buoyed by the recent renaissance of the latter as an attractive and highly proactive research field.

It is vitally important in assessing the role of GIS to mention, albeit in passing, the dominant theoretical underpinnings of mainstream regional survey. This not only serves to highlight the more immediate conceptual similarities behind the two approaches, but, more importantly, it leads on to one of the main arguments of the present discussion. This is to assert that the ease with which GIS has been incorporated into the methodological remit of survey, and the role it has been assigned, are largely due to a very particular theoretical conceptualization of what exactly GIS *is*.

A popular strategy within regional survey is to portray GIS as a spatial toolbox. This is seen as both atheoretical and universal in its applicability. Survey projects produce spatial information — the GIS handles spatial information. Underlying this is an understanding of GIS that stresses its neutrality, little more than a catalyst that speeds up the process of survey without playing an active part (cf. Wheatley 1993: 134). GIS as *effective* technology. This is perhaps rather ironic given the tendency amongst archaeologists to see GIS not in terms of 'acronym-as-noun' but more 'acronym-as-verb'. GIS as an *affective* tool that is somehow going to alter or change the information fed into it. Hence the apocryphal, yet oft-heard statement that a project is going to 'GIS' its data.

It should be noted that archaeology is not unique in being undecided as to what precisely GIS is. Within disciplines like Geography, debates rage as to whether GIS represents a simple tool or a science in its own right (Wright *et al.* 1997; Pickles 1997), and whether GIS simply speeds up existing tasks or fundamentally changes their context (Martin 1996: 258, as highlighted in Witcher, in press). Looking to the broader issue of perceived neutrality, and the impact of such technologies as GIS upon the lifeworld, critiques of such a perspective are well established within the field of social theory. The general thrust of these criticisms can be summed up in a quote from the philosopher Ihde, who has argued forcibly that all technologies should be regarded 'as "cultural instruments" which are non-neutral and deeply embedded in daily life praxes' (Ihde 1993: 13).

Leaving such detailed and often esoteric debates aside, what this dominant archaeological conceptualization of GIS-as-tool fails explicitly to acknowledge is the fact that the generic GIS is far from neutral. Even the most basic of systems carry an enormous amount of theoretical baggage (Pickles 1995; Gillings 1996). This can be highly generic, a good example being the realization and rigid enforcement of a very particular notion of space as a universal, quantifiable and profoundly external backdrop. This understanding of space is encoded within the underlying data models of the systems we use, such as the unambiguous point, lines and areas of vector systems, and may owe its origin more to the requirements of North American urban and rural planning than to any real world spatial phenomena. The embedded theory can also be far more subtle, a good example being the default viewpoint adopted by the GIS, which presents, in effect, everything from nowhere. It can be argued that this divorced viewing platform owes much to the military ancestry of many systems and the requirements of covert surveillance, and serves to conceal a number of subtle power relations within its ostensibly objective perspectival stance (Gregory 1994: 65).

The point being made here is that even at its most tool-like, GIS is never neutral. Much in the same way that data derived from a given programme of surface survey is rich with the theory that drove and shaped the collection, the generic GIS carries a body of implicit theory encoded within it. Just as a survey archive does not represent a neutral collection of objective facts, neither does the GIS represent a neutral collection of simple tools with which to study them. I intend to argue that this understanding of GIS as neutral, unproblematic tool, has served until very recently to restrict the utilization of GIS to a number of set caricature roles, such as the organization and management of data and production of static distribution maps. The implications that arise from directly challenging this conceptualization will be developed when we come to discuss the future of GIS within the survey context. Before we reach that stage it is important to look at the regional survey background and how GIS has tended to be applied up until now.

The survey context

Methodologies

The developmental history of surface survey as an investigative procedure can be characterized by a trajectory of near-continual methodological refinement, painstakingly dealing with what one group of researchers have termed 'pragmatic and procedural issues' (Alcock *et al.* 1994: 137). This is not to say that significant debates and critical methodological innovations have not taken place as a result of this introspection. As a result of this emphasis, practitioners have grown accustomed to frequently and reflexively re-evaluating their methodologies (e.g. Bintliff and Snodgrass 1985). A good example is the recognition and elevation of off-site archaeology to the status of a legitimate area of study in its own right (Gaffney *et al.* 1985; Bintliff and Snodgrass 1988; Cherry *et al.* 1991). In the light of this, the adoption of GIS can be seen as the continuation of a trend that has helped to embrace a range of 'tools' over the years, from geophysical survey to the detailed analysis of soil chemistry (e.g. Rimmington, in press). As mentioned in the introduction, the result of this dominant methodological focus has been the critical formulation of a range of techniques that has facilitated the careful recording of a wealth of information, information that may be ideally suited to examine changing cultural landscapes over significant tracts of social space and time. That this is rarely being exploited during the course of routine synthesis, analysis and interpretation can be attributed to a number of factors. Most dominant are those inexorably linked to the enormous breadth and scope of the data under study.

Data complexity

In the first instance we have the inherent complexity of the collected data sets themselves. This relates not only to artefact-specific information but also to metadata issues. The latter is a broad category, comprising in effect data about data. It encompasses a variety of issues, from sampling fraction and collection strategy through to post-depositional, environmental and macro-economic considerations, such as residuality and variations in supply (for a detailed discussion of such factors see Bintliff, Terrenato, Kuna, Barford *et al.*, this volume). All of these factors may have served to shape our carefully mapped spreads of cultural material. This complexity is in turn further compounded by the highly dynamic composition of these data sets, continually growing as the results of specialist studies and analyses are fed back into the developing project database.

Running alongside this is a second issue related to the strategic position of surface survey within landscape research. Following the pattern established by such pioneering regional survey projects as the UMME (McDonald and Rapp 1972) and southwest Argolid (Jameson *et al.* 1994), there has been the growing contextualization of such intensive studies within the body of broader multidisciplinary regional research frameworks. As a result, surface survey frequently finds itself integrated within a much wider body of specialist studies, generating very diverse data sets in their attempts to illuminate and access facets of the past landscape (for a recent example see Mee and

Forbes 1997). The fact that this is often undertaken and presented in a rather piecemeal fashion owes much to its origins in systems theory, a point I will return to later.

Interpretation and analysis

In highlighting the dramatic advances that have been made in the practical mechanics of survey, it must be noted that there has not been a similar advance in the development of the conceptual mechanisms required for the effective study, analysis and interpretation of the enormous volumes of information yielded by the cross-disciplinary and specialist-intensive programmes of study. Put simply, as the complexity of the database has grown through adherence to a continual agenda of methodological development and innovation, the conceptual mechanisms required for archaeologists to be able to articulate and explore the full range of information encoded within their data sets have not. This is despite the repeated efforts of a number of researchers, most notably Schofield (e.g. 1990), to foreground the issue.

As a result we too often see a mode of interpretation characterized by the extreme reduction of the meticulously recovered and carefully recorded micro-details and complexities essential to a fuller understanding of the people-place relations under study. Characteristic of such approaches is the rendering down of the culturally 'thick' primary survey data into aseptic, period-specific distribution maps. It could be argued that such depictions, distilled from the wealth of information collected during survey, illustrate at best broad trends that may be of questionable relevance to the often detailed questions that prompted the analysis. More dangerously, these dot-plots may take on the status of foundational data layers, adopting the mantle of primary data resource from which interpretation proceeds. In addition to this tendency towards generalization and reduction, the variety of disciplinary components central to the tenets of regional survey often find themselves forming discrete chapters in the final monograph. Rarely are their details fully integrated into the synthesis and discussion, in effect leading to the 'inter' as opposed to multidisciplinary approach as discussed critically by Jameson *et al.* (1994: 150).

In moving from methodological considerations to the business of interpretation and extracting meaning, we find ourselves facing a similar paradox to that encountered when sampling. It is only *after* the major programmes of regional survey have been completed that the full dynamic complexity and highly provisional nature of the data sets collected can be appreciated. However, by this stage it is too late to modify and focus the field methodologies and theoretically guided survey rationales used to acquire them. Acting

as a further interpretive constraint, surveys originally set up to address problems that may have originated in the context of a systems-based approach, with the environment portrayed as a proactive determinant, may now find themselves concerned with questions that have more to do with *mentalités*, symbolic structures and hermeneutic phenomenology than simple site catchment analyses.

It is fair to say that researchers are only now beginning to grasp the enormity of the task (Bintliff, in press). Previously taken-for-granted approaches are being challenged and questioned for the first time, such as the generation of normative rules of 'site' definition and the anthropogenic mechanisms behind such useful heuristics as the manuring hypothesis (Bintliff, in press; Alcock *et al.* 1994). In addition, the question of scale has been brought to the fore, challenging the primacy of the notional 'region' through a foregrounding of the term itself (Fotiadis 1997); a growing realization of the importance of more particularist scales of study (e.g. Gaffney *et al.* 1985; Whitelaw 1991); and at the other extreme the need for some degree of wide-scale inter-survey comparability (e.g. Alcock 1994; Bintliff 1997a). The importance of the latter has been given further impetus by the requirements of resource management and evaluation, with programmes for the standardized national reporting of surveys being widely advocated, for example the scoring schema proposed by Schofield in the context of lithic scatters (Schofield 1993).

Rather than concentrating upon the development of ever more efficient means of producing static map sheets, if GIS is to live up to early expectations, these are the challenges that it must face up to:

1) growing data complexity;
2) the importance of metadata;
3) the synthesis and integration of component disciplinary analyses;
4) the incorporation rather than reduction of 'thick' detail;
5) a variety of theoretical perspectives;
6) a decentring of concepts such as 'region' and 'site'.

Before going on to consider how these challenges might be met it is useful to look firstly at how GIS has impacted upon survey to date.

The current role of GIS

Preamble

> 'In either case we [the editors] believe that the combination of landscape archaeology and GIS is one of the most profound and stimulating combinations in archaeological theory and method in the 20th century' (Green 1990a: 5).

As intimated earlier, the widespread adoption of GIS within archaeology took place from the early 1990s. By contrast, most of the major regional survey programmes instrumental in shaping current practice had their origins in the 1970s and early 1980s. Application of GIS can therefore be seen as a largely retroactive process (Gillings and Sbonias, in press) in so far as the data under study had already been collected and, invariably, some preliminary and often highly implicit level of analysis had already been undertaken before the data were integrated into a GIS framework. As has been noted by Lock *et al.* (in press), this most commonly took the form of the identification and delineation of discrete activity foci or 'sites'. As was discussed earlier, the reason that the GIS could so easily be incorporated within the structure of survey programmes was due partly to the underlying conceptual similarities between regional survey and GIS; partly to the continual agenda of methodological development and evolution that has been the hallmark of regional survey; and finally to the fact that, by depicting GIS as an atheoretical tool, its application was seen as unproblematic. The net result of these trends has been a marked level of misunderstanding among survey archaeologists as to what precisely GIS represents. Too often it has been portrayed as little more than a highly sophisticated cartographic engine, a rapid and consistent method for the production of distribution maps. As a result it has been used extensively to do precisely that. GIS as *effective* tool. Whilst there is nothing inherently *wrong* in using a GIS to generate distribution plots (and, as Fisher has recently noted, practitioners should not be ashamed in owning up to such [Fisher, in press]), it can be argued that it is at the expense of the very innovative functionality that makes the systems so unique. As I hope to illustrate, the dominant definition of GIS within the survey literature describes better a developing group of computer-based approaches called Desktop Mapping Systems (DMS). A definition and suggested structure for the use of these highly complementary technologies will be explored in the final section.

How has GIS been applied in the context of methodology?

Where GIS has been adopted, its main methodological impact has been within the realm of data integration, management and curation. The thematic relational structure of the spatial and attribute data models makes the task of bringing together disparate sets of data into a single uniform environment relatively straightforward. This structure also makes the task of data management conceptually clearer, while the ability of GIS to work with data at a variety of scales has also proved critical. In effect, GIS provides survey projects with a uniform framework within which to bring together and curate data derived from a varied range of sources.

Many projects now make routine use of GIS in this context, for example, in the case of the Boeotia project GIS has been used to structure and manage the recovered archaeological database. Since the adoption of GIS in 1991, efforts have been directed solely towards the integration of the enormous archaeological data archive (Gillings and Sbonias, in press). This has involved the retrospective processing of data collected over 12 seasons of survey and detailed study, comprising some 10,000 macro-scale off-site transect units and more than 200 discrete, detailed site surveys. As specialist artefact reports, micro-topographic and geophysical surveys are completed and processed they are incorporated seamlessly into this single management environment.

More commonly, GIS is used to integrate partially interpreted archaeological data, in the form of site distribution maps, with a range of primary environmental and topographic base-maps. Typically a range of maps and satellite images summarising the foundational environmental variables for the region under study are integrated within a single spatial data model. Examples of this type of application abound in the literature (e.g. Gaffney and Stancic 1991; Baena *et al.* 1995; Gillings 1995).

It is in this area, in effect one of 'meta-archaeology', that the role of GIS can be seen as least problematical, as it is in the context of data management that GIS can be most effectively thought of as a facilitating tool. The success GIS has achieved in this sphere of efficient storage and data management can be most readily evinced by the enthusiasm with which national archaeological agencies responsible for the management of systematic and non-systematic survey-derived datasets are embracing the technology (e.g. Brandt *et al.* 1992; Guillot and Leroy 1995; van Leusen 1995). Despite such widespread acceptance, it should be noted that, even in its most tool-like manifestation, there still exist a number of shortcomings, intimately related to the underlying conceptual structures of GIS. Foremost amongst these is the issue of enforced standardization. When primary survey data are integrated into the relational spatial and attribute databases of GIS, an inevitable degree of reduction occurs. As intimated earlier, this can involve spatial compromise, where spatial information is shoehorned into a point, line or area category, or is generalized within a raster grid. It can also frequently involve attribute compromise, resulting from the distillation of the often highly generalized and qualitative survey records into the rigid structures of a formal database (this is not only a problem with computer-based systems, see Barford *et al.*, this volume). A common example is the generalization of detailed period information into a number of broad chronological phases.

The integration of environmental data also often involves some degree of reduction, such as the routine simplification of complex soil and geology maps into more manageable numbers of classes (e.g. Gaffney and Stančič 1991: 38; cf. Hunt 1992: 284). The important point to be reiterated here is that in no sense can GIS be regarded as theoretically neutral, and accusatory charges, such as that of technological determinism, a term that will be discussed in more detail shortly, can even be raised in the context of this most 'tool-like' application of the technology.

In terms of the impact upon practical field methodologies, the largely retrospective nature of the application of much GIS, coupled with the dominant notion of GIS as a largely passive tool, have served to restrict the implications for fieldwork to secondary issues. These are principally concerned with ease of data integration and the importance of applying rigour to the recording of spatial information, in effect speeding up the process of getting data into the system. For example, in the case of the Boeotia project, a largely but not exclusively retrospective application of GIS, the sole impact upon field methodology was the introduction of hand-held computers both to standardize the field recording and to speed up the process of data integration (Gillings and Sbonias, in press). This was coupled with the endorsement of an imposed, regularly gridded transect system to enable the spatial sampling units to be more rapidly and accurately defined in the GIS.

As an aside, it is interesting to note that, even when restricted to such secondary issues, there is no clear consensus amongst practitioners as to the precise impact GIS has had upon field practice. This factor is illustrated by a recent debate where researchers working on the Sangro Valley survey, one of the few new survey projects designed from the outset around an integrated GIS approach, advocated precisely the opposite tactic to that of the Boeotia survey when faced with the regular versus irregular grid issue (Gillings and Sbonias; cf. Lock *et al.*, in press)!

How has GIS been applied in the context of analysis?

Given the almost zealous optimism of the first wave of practitioners, it is in the field of analysis and interpretation that we can reasonably expect GIS to have had its most marked impact. Yet it is in precisely this realm that GIS can be said to have failed most convincingly. An oft-heard criticism, aired both by those directly involved with GIS and those unfamiliar with the technology, is the accusation that GIS-based analyses in archaeology have comprised little more than a range of applications the GIS could already routinely perform. This tendency has been elegantly referred to as 'technological determinism' (Harris and

Lock 1995: 355). In any study involving spatial data there is the issue of *needs*, in our case the archaeological problematic, versus *constraints*, with the possibility that the former will have to be seriously modified in light of the latter (Association for Geographic Information 1996). These constraints can be numerous, ranging far wider than the functional abilities of the systems employed to embrace such issues as data availability and reliability through to cost and compatibility. To give a rather coarse archaeological example, a standard soil map is much easier to locate and incorporate than a map summarizing symbolic pathways and socially embedded places present in a monumental prehistoric landscape. The tendency within archaeological applications of GIS until now has been to privilege constraints over needs, with the archaeological problematic invariably compromised or simplified. A good example is the fact that often modern environmental maps, while easy to obtain and convenient to integrate into a GIS database, may be of limited relevance to the periods under study. Yet while this fact is acknowledged, they are used within GIS-based studies anyway, to the detriment of the archaeological analysis being attempted (e.g. Verhagen *et al.* 1995: 200–201; cf. Gillings 1995; 1997a; 1997b).

This sacrifice of needs to constraints has resulted in an analytical contribution, which can best be described as stagnant and sterile, dominated by simple distribution plotting, map overlay and basic statistical description. If these trends can be characterized, it is by an almost formulaic requirement to treat sites as discrete point locations, and then examine them with respect to a list of factors so familiar to anyone who has read the GIS literature as to form a credo: slope, aspect, soil type, hydrology, etc. That this runs the risk of becoming an uncritical orthodoxy can be illustrated with reference to a recent discussion where the decision to pursue such a strategy was justified solely on the grounds that such factors 'constitute a suite of variables that is now a traditional element of locational analysis in archaeology' (Martlew 1996:293). The issue has been elegantly summarized by Wansleben and Verhart, who assert that in such cases it is the ease with which researchers are able to perform such measurements that makes them simply too tempting to resist (1997: 57). The danger is that the requirements of the archaeological problem will too often give way to a combination of lowest common denominators, what the GIS does best coupled with the data that are easiest to obtain.

Far from shaping the GIS environment to incorporate contemporary developments in theory and practice and help build towards a fuller understanding of the problem at hand, the abilities of the GIS effortlessly to manipulate spatial data have led instead to a renewed interest in spatial-analytical techniques and environmentally dominant approaches championed in the early 1970s but largely since ignored (a factor noted by Wheatley as early as 1993; for a recent discussion see Bintliff 1997b: 70–72). This is illustrated by the resurgence of interest in such approaches as Thiessen polygons and the uncritical application of simple site catchment analysis within survey projects (e.g. Savage 1990; Hunt 1992), despite the existence of critiques within the spatial analysis literature contemporary with the first wave of such applications (e.g. Dennell 1980). Ironically, in the case of the longer-lived regional survey projects, the resulting analyses may well fit comfortably within the remits of the early project design.

One trend has been the enormous upsurge in predictive modelling exercises. This has been undertaken in the context of resource management (e.g. Judge and Sebastian 1988; Altschul 1990), where the protection and curation of archaeological remains is paramount. It has also been used to formulate and test models in the past, most commonly working inductively from the known locations of sites in surveyed areas to predict the locations of comparable sites in areas not covered by survey (e.g. Gaffney and Stancic 1991: 66–76; Dann and Yerkes 1994: 304, predicting Roman villa and Frankish tower locations respectively). Although a critical discussion as to the merits or otherwise of such approaches is beyond the scope of the present paper, it should be noted that for most researchers, given the enormous developments in theory and more critical understandings of space, place and landscape, such applications are of limited value. As Hodder has remarked, '[I]t is possible to predict many aspects of human behaviour with some accuracy but without any understanding of the causal relationships involved' (Hodder 1992: 100). As many landscape based researchers are aware, the adage that prediction has little to do with explanation takes on added significance in the context of GIS.

Summary: from black box to Pandora's box

There is always the danger that in attempting to characterize the broad adoption and application of GIS within regional survey the picture painted is simplistic. There are certainly exceptions to the situation I have highlighted above, with a growing number of researchers beginning to take the first steps towards the realization of GIS as more than simply a tool (e.g. Wheatley 1993; Gaffney *et al.* 1995; Gillings and Goodrick 1996; Llobera 1996; Gillings 1998). It is fair to say that such approaches are currently few and far between. The dominant definition of GIS within survey remains that of a neutral tool. Moreover, a tool designed to produce maps. Far from involving GIS *actively* in the process of interpretation and analysis, it has often instead served as little more than

a window into the project database. As a result, the realization of GIS as omnipotent black box and heuristic panacea confidently predicted by the first wave of practitioners has been replaced by a sense of GIS as Pandora's box, brimful with regressive theory and technologically deterministic practice.

Future practice

GIS and survey

The critical point to emphasize is that this sense of stagnation and stasis is a result of the combined tendencies to treat GIS as a neutral tool and tailor questions and analyses around its capabilities. That this does not have to be the case is evinced by the growing number of more informed and reflexive applications of the technology. The success of such developments is heavily reliant upon the undertaking of a fundamental reconceptualization of GIS within the regional survey context. Two alternatives exist. The first is to develop a theory of practice for GIS unique to archaeology in general and to the more specific case of regional survey. This would serve to address the challenge of developing theoretical frameworks, whether Annales, structuralist or phenomenological, privileging needs over constraints and developing a notion of GIS as *affective* approach. The other alternative is to persist with a notion of GIS as neutral catalyst tool, effectively closing a loop back to the highly functionalist agendas characteristic of the more quantitative climate that gave birth to regional survey as we understand it today. That GIS has an enormous amount to offer regional survey is not in doubt. Needless to say, in suggesting a possible trajectory for the future development of GIS within survey it is the former option I wish to elaborate.

A critical reformulation of GIS

Central to this are two fundamental changes in the way we perceive and use GIS. The first is a reconceptualization of the status of GIS from neutral tool to theoretically informed approach. The second is to reformulate analyses involving GIS so that they are led by the archaeological problematic and not by constraints. It is encouraging to note that within archaeological–GIS research, such a move is already under way, with the reformulation of GIS interpreted in a variety of ways. Some simply envisage GIS as a single integrated component in a much broader Archaeological Information System (AIS), itself part of a more generic Archaeological Knowledge System (AKS). This is the most radical and fundamental reconceptualization, with the very existence of GIS as an independent research area brought into question (McGlade 1996; 1997). Others persist with the tool metaphor but challenge the direct equation of GIS with a single neutral toolbox. Instead a number of levels are identified, conforming not only to the more tool-like data management and simple mapping tasks that have characterized GIS up until now, but also to the testing and exploration of theoretical constructs (Gaffney and van Leusen 1996). The first of these trajectories is perhaps inevitable and is largely to be encouraged, GIS and related technologies receding to become part of the fabric and context of survey, assumed yet not explicitly highlighted. The second breakdown is useful, but by privileging the highly problematic term 'tool', loaded as it is with implications of neutrality, it runs the risk that much of its potential impact could be misconstrued or even negated. As a result the term 'approach' is adopted here. Rather than advocating the use of a GIS we should instead be advocating the adoption of a GIS approach. This is not simply a matter of syntax. The first encourages the application of a neutral tool to address a specific problem or failing that prompt a reformulation of the problem. The second sees GIS as serving not only the tool-like functions that have so effectively characterized its application to date, but also helping to create or clear a flexible and dynamic conceptual-analytical space. This is the creation of an environment within which not only to manage and plot information, but also to integrate, explore, undertake analyses and help furnish explanations. It is as much a realm for the articulation of ideas as for the production of distribution maps.

GIS and the rise of desktop mapping applications

Moving away from such broad conceptual matters, it is important to consider how this notion of GIS as an exploratory and interpretive environment is to be implemented at the more practical level. Within regional survey a situation can be envisaged where projects utilise both GIS and DMS (Data Management Subset), with the latter taking on the role traditionally assigned within surveys to GIS. DMS can be thought of as a cartographic and attribute database management subset of the GIS, relying heavily upon the metaphor of the map to both manage data and allow users to interact with it. The interface between the data and the user is realized via the simple and familiar medium of the map.[1] Whilst such systems lack the analytical capabilities and flexibility of the true GIS, they have more than sufficient functionality for the majority of survey management and data exploration work and, perhaps more importantly, have a friendly and intuitive user interface. It should be noted that the latter can also be construed as a potential weakness, with the fear that the ease-of-use will encourage a commensurate sense of black box functionality. This could serve to mask the fact

that for all its push-button efficiency, the DMS carries the same underlying theoretical baggage as the GIS from which it is derived.

Within the GIS framework outlined here, DMS will take on the majority of mundane tasks previously allocated to the GIS, enabling researchers to tailor and maximize the unique functionality of the latter. The point to emphasise is that the adoption of DM does not signal the end of GIS in regional survey, if anything it serves to define more clearly its true role and position. This is best illustrated through two brief examples.

Looking to the archive of data generated by the Boeotia project, one site that has provoked enormous interest is the Frankish and Early Turkish period settlement of VM4, located adjacent to the Classical town of Askra in the Valley of the Muses (both discussed in their historic context in Bintliff 1996). The ceramic assemblage derived from the survey of this settlement has enormous implications for the generation of a detailed ceramic typology for this long-neglected period of Greek history. In addition, detailed textual records pertaining to the site were discovered during recent work amongst the contemporaneous Ottoman archives, making it an ideal candidate for the development of the kind of structural historical approach characteristic of the more Annales-informed school of interpretation and analysis (for further discussion of the site VM4 see Bintliff, this volume).

In discussing the relationship between GIS and DM let us look firstly at the role the GIS is playing in the analysis of this complex site. In its most straightforward application, the GIS has been used to integrate the off-site and site transect information, converting annotated field maps and notebook entries into a set of carefully constructed and topologically consistent transect layers. Once constructed, these have been linked directly to the attribute information relating to the finds made within each of the transect units.

Moving beyond this relatively simple data management task, the GIS has also been used to process and explore a data set resulting from a highly detailed micro-topographical survey that was undertaken of the village environs. One of the key research questions relating to VM4 concerns the location and form of the numerous house structures that would have been present at the site. As a result of post-abandonment robbing for the probable construction of sheep-folds and field boundaries, compounded by a dense mat of prickly oak that obscures large areas of the site environs, domestic structural remains appear to be nonexistent. The micro-topographical data, a quasi-regular grid of three-dimensional point coordinates, was collected in the hope of capturing any surviving traces of the habitation areas surviving beneath this dense carpet of vegetation (Figure 9.1).

Once the data had been collected and collated, the GIS package Arc/Info was used to construct a series of digital elevation models, that is, continuous representations of the surface topography. Using the surface analysis functionality of the GIS these were then care-fully interrogated, revealing clear groups of previously unsuspected house foundations (Figures 9.2a–9.2c).

The application of DM within the VM4 study has been to take the results of this analysis and com-bine them within a single, highly flexible interpretive

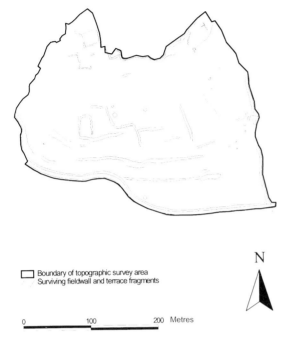

Boundary of topographic survey area
Surviving fieldwall and terrace fragments

N

0   100   200   Metres

Figure 9.1   The extent of the topographical survey area, showing the positions of extant wall and terrace features visible on the ground surface (for a general location map of the site in relation to the Valley of the Muses see

Figure 9.2b The DEM processed within the GIS to derive a grey-scale map showing the slope of each portion of the landscape. These range from shallow (dark grey) through to very steep (white). In this image not only are the surviving terrace and wall fragments indicated on Figure 9.1 visible, but also a cluster of small sub-rectangular features in the south-west block of the survey area.

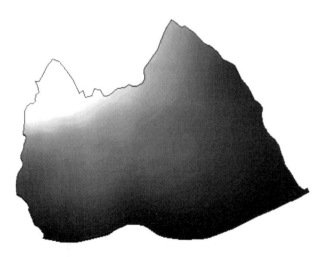

Figure 9.2a The GIS-derived Digital Elevation Model (DEM). This surface model was generated on the basis of over 8000 measured survey points. In the DEM elevations range from low (black) through to high (white). As can be seen, the general sloping nature of the hillside is well represented but no micro-topographical detail is visible.

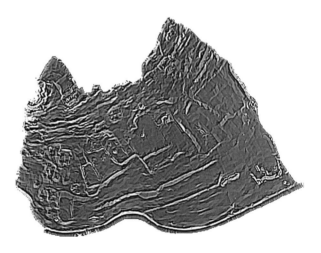

Figure 9.2c In this image the GIS has been used to enhance the edges of these features through careful filtering of the slope map. The sub-rectangular features can now be clearly identified as house foundations. Additional structural remains, for example, the small cluster of foundations in the central-northern portion of the survey area, can also be identified.

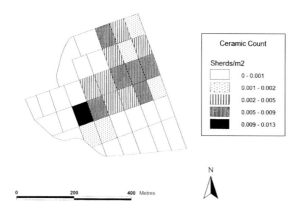

Figure 9.3a A total ceramic density plot for the grab sample taken from the site during surface survey. The survey grid and related database of artefactual information have been displayed using a Desktop Mapping system.

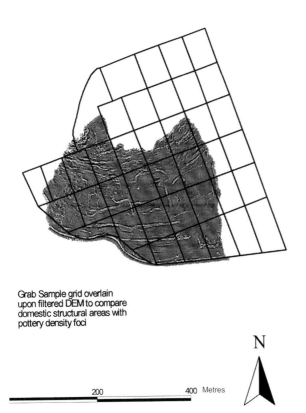

Grab Sample grid overlain upon filtered DEM to compare domestic structural areas with pottery density foci

Figure 9.3b By using the Desktop Mapping system, the results of the micro-topographical survey can easily be integrated with the surface survey database. Here the grab sample grid and survey results have been integrated to enable the identified density patterns to be related to the domestic structural remains. Interestingly the main densities of artefactual material occur away from the identified structures.

environment with maps encoding the results of geophysical and soil chemistry studies; data layers recording the positions of clearance cairns and relict field walls; and the survey derived ceramic density maps (Figures 9.3a and 9.3b).

A further example can be seen in recent research within the flood-plain of the river Tisza in north-east Hungary (Gillings 1998). In this study the GIS was used to furnish complex hydrological reconstructions and a set of 'fuzzy' simulations that encoded in a map-like form the uncertainty and ambiguity of given palaeo-flood events. The results of these analyses have been brought together, along with site data recovered through a programme of intensive surface survey, using a DM utility.

To summarize, in the case of a typical field survey project, the tasks of assembling and editing basic data layers, whether archaeological or otherwise, and the derivation of specific thematic layers from them, are best undertaken using the powerful capabilities of GIS. The exploration of these layers with the complex spatial and attribute databases generated by field survey are tasks ideally suited to the DMS. Although separated here for the purposes of clarity, in practice the relationship between GIS and DM is one of complementarity and synergy.

The impact upon practical methodology

As discussed earlier, the impact upon field methodology has been limited, restricted largely to issues concerned with optimizing data input. I intend to argue here that, in the context of regional survey, this minimization of impact is a trend that is to be encouraged in the future rather than rectified. The development of field survey methodologies have been characterized by their reflexivity and problem-driven nature. Techniques have been formulated to answer very specific archaeological questions and have been optimized and enhanced as a result of the data obtained. By explicitly modifying fieldwork methods around the GIS, whether by forcing ever more reduction and standardization in the recording of attributes or through the regularising of transect layout, a practice that may make little sense when considering issues such as surface vegetation cover and fieldwork time restrictions, we run the risk once again of elevating technological constraints over archaeological needs.

In methodological terms the real benefits of the GIS approach will not come from its utility in the arena of everyday fieldwork. Instead they will come from its ability to serve as a test-bed for the development and evaluation of new approaches, and through its ability to realize existing epistemological goals, for example, the creation of a truly inter-, as opposed to multi-disciplinary project framework. The GIS environment is also ideally suited to explore problem areas that have persisted despite three decades of productive debate: for instance, it can be used routinely and rapidly to assess the capabilities and performance of a host of alternate sampling strategies in a range of conditions, both real and simulated. It can also provide a useful mechanism for the study of the various models put forward over the years to explain such phenomena as the manuring hypothesis, and the effects of visibility and walker biases upon the collected sample (for a discussion of such factors see Terrenato and Kuna, this volume).

One highly active research area concerns the field of interpolation, to be distinguished from predictive modelling, which is concerned with the problems associated with extending trends identified within survey samples to whole study areas. Here pioneering theoretical work into optimum interpolation strategies and the effects of transect shape, frequency and orientation (Robinson and Zubrow, in press) along with a number of practical statistical case-studies by Wheatley (1996), have enormous implications for the subdiscipline of surface survey as a whole.

Looking to the issue of interdisciplinary integration, a number of studies have already served to demonstrate the utility of the GIS approach in bringing the initially diverse results generated by discrete thematic studies together into a single analytical environment. In the Upper Tisza project, an intensive survey undertaken in the far northeast of Hungary, GIS has provided the medium within which to integrate topographical, geomorphological, hydrological and survey derived data in an attempt to recreate a flooded Neolithic landscape (Gillings 1995; 1997a; 1997b; 1998). Another example can be seen in the recent work of Verhagen (1996), modelling soil erosion potentials.

The final area I intend to mention is that of scale. Regional survey has always functioned broadly at two scales, that of the general landscape and that of the site. However, with the publication of many of the major survey projects there has come the growing realization that regional survey tends to identify more problems than it answers. This has led to a growing trend amongst practitioners to view intensive regional survey as constituting a first stage, macro-level research strategy, ideally to be followed up by a much more problem-targeted and particularist programme of micro-survey (e.g. Chapman *et al.* 1997). The ability of the GIS approach to integrate data at a variety of scales makes it an ideal environment within which to formulate and explore such methodological developments.

The impact upon analysis and interpretation

In each of the cases outlined above, as well as acting as a tool and catalyst, the GIS serves to enrich the interpretive environment. It has enormous potential to realize not only existing goals and approaches but also to develop new ones. A good example can be seen in the context of phenomenology, a philosophical perspective that has found increasing popularity among landscape archaeologists. Here the importance of such issues as everyday experience, encounter, social practice, perception and embodiment are stressed. Although phenomenology has been criticized as something one preaches more easily than practices (Simmons 1993: 79), a number of new, innovative, GIS-based studies involving such techniques as enhanced visibility analysis, virtual reality (VR) modelling and the development of complex, integrated time-space models, have begun to explore the routine practical application of such issues (Gillings and Goodrick 1996; Llobera 1996; Gosden and Lock 1996; Gillings 1998). Needless to say, these developments have enormous significance for the study and interpretation of survey-derived data.

To give an indication as to how these more experiential theoretical perspectives may be integrated into the survey environment and serve to enrich our understandings, we can turn again to the Boeotia project database. An example of precisely this kind of approach is currently being formulated for the study of the hinterland of the Classical city of Hyettos, the project control area located to the north of the core study zone. Here one of the directors of the project (John Bintliff) noticed that classical farmsteads in the surrounding countryside, although severed from the acropolis by physical distance, were still ideally sited visually to overlook the cult activity taking place upon it. In addition, the positions taken by these sites took advantage of the unique acoustic effects of the basin 'auditorium' surrounding the acropolis and lower city, enabling them to partake aurally as well as visually in the activities taking place. The implication is that, although separated from the ritual and social life of the city in quantitative spatial terms, the occupants retained an immediate sensual attachment. Work is currently under way to develop GIS-based models of the propagation through the landscape of the low-fidelity sounds generated by the hubbub of the lower city and more punctuative high-fidelity sounds associated with the cult activity atop the acropolis. In the first instance the aim of the GIS study is to generate a number of mapped soundscapes, dynamic 'murmur maps' and highly sensitive intra-plateau viewsheds that can then be integrated into a single exploratory-analytical environment using a DMS. In the next stage, the GIS will play an important role in the generation of interactive, sound-enriched VR models and explorations of alternative and more sensual metrics of space.

Further examples of such approaches can be seen in the work of Belcher *et al.* (in press) in their study of tomb locations, and in the fascinating work into the translation of symbolic schema into physical spatial form being undertaken in the cognitive investigations of Zubrow (1994).

Publication and dissemination

I would like to end by looking at an area where GIS is going to have an increasing impact within regional survey. This concerns the effective and rapid publication and dissemination of data and results. Although undoubtedly important, the major contribution will not be in the more effective production of hard-copy distribution plots and maps. As has been stressed throughout the present discussion, the enormous complexity and highly dynamic nature of the data yielded by regional survey require an equally dynamic and flexible mechanism for their management and analysis. The GIS approach provides us with precisely this type of mechanism. That this analytical strength can become a weakness when we reach the stage of publication has been realized in practice by all those engaged in GIS research. In much the same way as via the medium of the DMS, the highly fluid way in which researchers engage with the complex set of project databases during the process of exploration and interpretation does not lend itself naturally to the production of static snapshot images. To give a simplistic example, a single site in the Boeotia database may have attribute information relating to a large number of discrete chronological phases and a host of form-functional types within each period category (see Gillings and Sbonias, in press). While this can be accessed and explored via the medium of a single, on-screen site transect map in the DMS, it would require an enormous, and potentially confusing number of repeated hard-copy plots to begin to communicate adequately the depth of this information.

To overcome the simplification and reduction inherent in a ream of static distribution maps or heavily generalized composite distribution map, a method of publication is needed that is complementary to the dynamics of both the data sets and analysis medium. A growing area of research within the field of GIS is in the linkage and interfacing of GIS and DM based environments directly to multimedia presentation software and, more excitingly, developments in the World Wide Web (WWW) capabilities of the Internet. Good examples are Ian Johnson's explicitly archaeological TimeMap project (Johnson, 1999) and the appearance of DM utilities like ArcExplorer, which are designed to interface directly with data via the WWW. The growth of refereed Internet-based journals, such as Internet Archaeology, and acceptance of CD-ROM-based multimedia as a legitimate form of publication will result in survey projects being in the

position to publish in an ideal format, free from lengthy textual gazetteers and either a huge number of distribution maps or heavily simplified reductionist summary plots.

A pioneering example of how such an approach can be put into practice can be seen in Sebastian Heath's excellent work on behalf of the Pylos Regional Archaeological Project (Davis *et al.* 1996) and Nemea Valley Archaeological Project (Cherry *et al.* 1996). Here detailed site-specific information held in the survey archives has been implemented as a fully interactive gazetteer that can be accessed and queried via the WWW using a common browser interface.[2] Although very much in its infancy, this is a significant development that deserves to be lauded, setting a standard which major regional survey programmes will undoubtedly follow. The fact that translation of these archives to a WWW-based format was a result of the intelligent adoption of a GIS-based approach, is no accident.

Conclusions

To conclude, GIS, coupled with developments in DM and information technology, are likely to play a vitally important role in the future development of regional surface survey. As the previous discussion has shown, this will impact at all levels, from methodology, analysis and the formulation of theory through to publication and communication. For this potential to be realized it is imperative that we cease to take GIS for granted, assuming its neutrality and ascribing it a number of restrictive predefined roles. Instead we need to begin to rethink our dominant conceptualization of GIS as little more than a tool.

In the preceding discussion an attempt has been made to examine the background context, identifying and attempting to explain the form this process of adoption has taken. As a result it has been possible to both highlight and question a number of dominant conceptual positions and prescribed roles, as a means of foregrounding the status of GIS not only as an effective tool but also as an affective approach.

In sketching out a possible trajectory for this realization of GIS, and I am sure that there are others, the approach taken has been to rely upon the highlighting of substantive issues and the use of broad illustrative examples. This is deliberate and in direct opposition to the advocacy of a single prescriptive framework, which carries the inherent risks of achieving little more than the establishment of a new orthodoxy. The aim instead has been to encourage attempts within regional survey to challenge and extend the allotted role of GIS, guided by developments in theory and the archaeological problematic at hand.

Acknowledgements

This paper was written as part of a research fellowship funded and supported by the Leverhulme Trust.

Notes

1) An example of a DM application already in common use among archaeologists is the Arcview program, produced by the company responsible for the industry standard Arc/Info GIS package. It is interesting to note that, as progressive versions of Arcview have begun to resemble a full GIS in terms of their functionality, the company has released a simpler DM application, Arc-Explorer, to fill in the gap. The free availability of the latter along with its ability to interface seamlessly with the World Wide Web will undoubtedly make it very popular among archeologists.

2) A textual description fails to do justice to these sites. Any interested readers with access to a Web browser are encouraged to explore them. The respective URLs are, for the Pylos survey, http://classics.lsa.umich.edu/PRAP.html, and, for the Nemea survey, http://classics.lsa.umich.edu/NVAP.html.

References

Alcock, S.E.
1994   Breaking up the Hellenistic world: survey and society. In I. Morris (ed.), *Classical Greece: Ancient Histories and Modern Archaeologies*, 171–90. Cambridge: Cambridge University Press.

Alcock, S.E., J.F. Cherry and J.L. Davis
1994   Intensive survey, agricultural practice and the classical landscape of Greece. In I. Morris (ed.), *Classical Greece: Ancient Histories and Modern Archaeologies*, 137–70. Cambridge: Cambridge University Press.

Allen, K.M.S., S.W. Green and E.B.W. Zubrow
1990   *Interpreting Space: GIS and Archaeology*. London: Taylor & Francis.

Altschul, J.H.
1990   Red flag models: the use of modeling in management contexts. In K.M.S. Allen, S.W. Green and E.B.W. Zubrow (eds.), *Interpreting Space: GIS and Archaeology*, 226–38. London: Taylor & Francis.

Association for Geographic Information
1996   *Guidelines for geographic information content and quality*. London: AGI.

Baena, J., C. Blasco and V. Recuero
1995   The spatial analysis of Bell Beaker sites in the Madrid region of Spain. In G. Lock and Z. Stančič (eds.), *The Impact of Geographic Information Systems on Archaeology: A European Perspective*, 101–16. London: Taylor & Francis.

Barker, G.W.W. (ed.)
1995   *A Mediterranean Valley: Landscape Archaeology and Annales History in the Biferno Valley*. Leicester: Leicester University Press.

Belcher, M., A. Harrison and S. Stoddart
in press Analysing Rome's hinterland. In M. Gillings, D. Mattingly and J. van Dalen (eds.), *Geographical Information Systems and Landscape Archaeology*. Mediterranean Landscape Archaeology 3. Oxford: Oxbow.

Bintliff, J.L. (ed.)
1991a   *The Annales School and Archaeology*. Leicester: Leicester University Press.
1991b   The contribution of an Annaliste/structural history approach to archaeology. In J.L. Bintliff (ed.), *The Annales School and Archaeology*, 1–33. Leicester: Leicester University Press.
1996   The archaeological survey of the Valley of the Muses and its significance for Boeotian history. In A. Hurst and A. Schachter (eds.), *La Montagne des Muses*, 193–210. Geneva: Librairie Droz.
1997a   Regional Survey, demography, and the rise of complex societies in the Ancient Aegean: coreperiphery, Neo-Malthusian, and other interpretive models. *Journal of Field Archaeology* 24(1): 1–38.
1997b   Catastrophe, chaos and complexity: the death, decay and rebirth of towns from Antiquity to today. *Journal of European Archaeology* 5(2): 67–90.
in press The concepts of 'site' and 'offsite' archaeology in surface artefact survey. In M. Pasquinucci and F. Trément (eds.), *Non-destructive Techniques Applied to Landscape Archaeology*. Mediterranean Landscape Archeology 4. Oxford: Oxbow.

Bintliff, J.L., and A.M. Snodgrass
1985   The Cambridge/Bradford Boeotian Expedition: the first four years. *Journal of Field Archaeology* 12: 123–61.
1988   Off-site pottery distributions: a regional and interregional perspective. *Current Anthropology* 29: 506–13.

Brandt, R., B.J. Groenewoudt and K.L. Kvamme
1992   An experiment in archaeological site location: modelling in the Netherlands using GIS techniques. *World Archaeology* 24(2): 268–82.

Bulliet, R.W.
1992   Annales and archaeology. In A.B. Knapp (ed.), *Archaeology, Annales, and Ethnohistory*, 131–34. Cambridge: Cambridge University Press.

Chapman, J.C., C.J. Pollard, D.G. Passmore and B.A.S. Davis
1997   Sites and palaeo-channels in the Polgar lowlands, north-east Hungary: the Upper Tisza Project 1996 field season. *Universities of Durham and Newcastle upon Tyne Archaeological Reports 20*, 12–21. Durham: University of Durham.

Cherry, J.F., J.L. Davis and E. Mantzourani
1991   *Landscape Archaeology as Long-Term History: Northern Keos in the Cycladic Islands from Earliest Settlement until Modern Times*. Los Angeles: University of California.
1996   *The Nemea Valley Archaeological Project Archaeological Survey: Internet Edition*, http://classics.lsa.umich.edu/ NVAP.html.

Clarke, D.L. (ed.)
1977   *Spatial Archaeology*. London: Academic Press.

Dann, M.A., and R.W. Yerkes
1994   Use of Geographic Information Systems for the spatial analysis of Frankish settlements in the Korinthia, Greece. In P.N. Kardulias (ed.), *Beyond the Site: Regional Studies in the Aegean Area*, 289–312. Maryland: University Press of America.

Davis, J.L., S.E. Alcock, J. Bennet, Y. Lolos, C.W. Shelmerdine and E. Zangger
1996   *The Pylos Regional Archaeological Project: Internet Edition*, http://classics.lsa.umich.edu/PRAP.html.

Dennell, R.
1980   The use, abuse and potential of site catchment analysis. In F.J. Findlow and J.E. Ericson (eds.), *Catchment Analysis: Essays on Prehistoric Resource Space*, 1–20. Anthropology UCLA 10.

Fisher, P.F.
in press Geographical Information Systems: today and tomorrow? In M. Gillings, D. Mattingly and J. van Dalen (eds.), *Geographical Information Systems and Landscape Archaeology*. Mediterranean Landscape Archaeology 3. Oxford: Oxbow.

Fotiadis, M.
1997   Cultural identity and regional archaeological projects: beyond ethical questions. *Archaeological Dialogues* 4(1): 102–13.

Gaffney, C., V. Gaffney and M. Tingle
1985   Settlement, economy or behaviour? Micro-regional land-use models and the interpretation of surface artefact patterns. In C. Haselgrove, M. Millett and A. Smith (eds.), *Archaeology from the Ploughsoil: Studies in the Collection and Interpretation of Field Survey Data*, 95–107. Sheffield: Department of Archaeology and Prehistory.

Gaffney, V. and Z. Stančič
1991   *GIS Approaches to Regional Analysis: A Case Study of the Island of Hvar*. Ljubljana: Research Institute for the Faculty of Arts and Science University of Ljubljana.

Gaffney, V., and M. van Leusen
1996   Extending GIS methods for regional archaeology: the Wroxeter Hinterland Project. In H. Kamermans and K. Fennema (eds.), *Interfacing the Past: Computer Applications and Quantitative Methods in Archaeology CAA95*, 297–305. Analecta Praehistorica Leidensia 28. Leiden: University of Leiden Press.

Gaffney, V., Z. Stančič and H. Watson
1995   Moving from catchments to cognition: tentative steps towards a larger archaeological context for GIS. *Scottish Archaeological Review* 9–10: 51–64.

Gibbon, G.
1989   *Explanation in Archaeology*. Oxford: Basil Blackwell.

Gillings, M.
1995   GIS and the Tisza flood-plain: landscape and settlement evolution in north-eastern Hungary. In G. Lock and Z. Stančič (eds.), *The Impact of Geographic Information Systems on Archaeology: A European Perspective*, 67–84. London: Taylor & Francis.
1996   Sounds stinky (but feels quite nice): towards a more sensuous GIS. Paper presented at the 18th Annual Theoretical Archaeology Group Conference, Liverpool.
1997a   Spatial Organisation in the Tisza flood-plain: Landscape Dynamics and GIS. In J.C. Chapman and P. Dolukhanov (eds.), *Landscapes in Flux: Central and Eastern Europe in Antiquity*, 163–80. Colloquia Pontica 3. Oxford: Oxbow.
1997b   Not drowning but waving? The Tisza flood-plain revisited. In M. North and I. Johnson (eds.), *Archaeological Applications of GIS: Proceedings of Colloquium II, UISPP XIIIth Congress*. Archaeological Methods Series 5. Sydney: Sydney University.
1998   Embracing uncertainty and challenging dualism in the GIS-based study of a palaeo-flood plain. *European Journal of Archaeology* 1(1): 117–44.

Gillings, M., and G.T. Goodrick
1996   Sensuous and reflexive GIS: exploring visualisation and VRML. *Internet Archaeology* 1 (http://intarch.ac.uk/journal/issue1/).

Gillings, M., and K. Sbonias
in press Regional survey and GIS: the Boeotia Project. In M. Gillings, D. Mattingly and J. van Dalen (eds.), *Geographical Information Systems and Landscape Archaeology*. Mediterranean Landscape Archaeology 3. Oxford: Oxbow.

Gosden, C., and G. Lock
1996   Emerging history: structure, landscape and GIS on the Ridgeway. Paper presented at the 18th Annual Theoretical Archaeology Group Conference, Liverpool.

Green, S.W.
1990a   Approaching archaeological space. In K.M.S. Allen, S.W. Green and E.B.W. Zubrow (eds.), *Interpreting Space: GIS and Archaeology*, 3–8. London: Taylor & Francis.
1990b   Settlement in south east Ireland: landscape archaeology and GIS. In K.M.S. Allen, S.W. Green and E.B.W. Zubrow (eds.), *Interpreting Space: GIS and Archaeology*, 356–63. London: Taylor & Francis.

Gregory, D.
1994   *Geographical Imaginations*. Oxford: Basil Blackwell.

Guillot, D., and G. Leroy
1995   The use of GIS for archaeological resource management in France: the SCALA project, with a case study in Picardie. In G. Lock and Z. Stančič (eds.), *Archaeology and Geographical Information Systems: A European Perspective*, 15–26. London: Taylor & Francis.

Harris, T.M., and G.R. Lock
1995   Toward an evaluation of GIS in European archaeology: the past, present and future of theory and applications. In G. Lock and Z. Stančič (eds.), *Archaeology and Geographical Information Systems: A European Perspective*, 349–66. London: Taylor & Francis.
1996   Multi-dimensional GIS: exploratory approaches to spatial and temporal relationships within archaeological stratigraphy. In H. Kamermans and K. Fennema (eds.), *Interfacing the Past: Computer Applications and Quantitative Methods in Archaeology CAA95*, 307–16. Analecta Praehistorica Leidensia 28. Leiden: University of Leiden Press.

Hodder, I.R.
1992   *Theory and Practice in Archaeology*. London: Routledge.

Hodder, I.R., and C.R. Orton
1976   *Spatial Analysis in Archaeology*. Cambridge: Cambridge University Press.

Hunt, E.D.
1992   Upgrading site-catchment analyses with the use of GIS: investigating the settlement patterns of horticulturalists. *World Archaeology* 24(2): 283–309.

Ihde, D.
1993   *Postphenomenology: essays in the postmodern context*. Illinois: Northwestern University Press.

Jameson, M.H., C.N. Runnels and T.H. van Andel
1994   *A Greek Countryside: The Southern Argolid from Prehistory to the Present Day*. California: Stanford University Press.

Johnson, I.

in press   Mapping the fourth dimension: the TimeMap project. In L. Dingwall, S. Exon, V. Gaffney, S. Laflin, M. van Leusen (eds.), *Archeology in the Age of the Internet: Computer Applications and Quantitative Methods in Archaeology 1997.* Oxford: Archaeopress (CD-ROM).

Judge, W.J., and L. Sebastian (eds.)

1988   *Quantifying the Present and Predicting the Past: Theory, Method and Application of Archaeological predictive Modeling.* Denver: Bureau of Land Management.

Kvamme, K.

1991   Preface. In V. Gaffney and Z. Stančič, *GIS approaches to Regional Analysis: A Case Study of the Island of Hvar*, 11–12. Ljubljana: Research Institute for the Faculty of Arts and Science, University of Ljubljana.

Last, J.

1995   The nature of history. In I. Hodder, M. Shanks, A. Alexandri, V. Buchli, J. Carman, J. Last and G. Lucas (eds.), *Interpreting Archaeology: Finding Meaning in the Past*, 141–57. London: Routledge.

Llobera, M.

1996   Exploring the topography of mind: GIS, social space and archaeology. *Antiquity* 70: 612–22.

Lock, G., T. Bell and J. Lloyd

in press   Towards a methodology for modelling surface survey data: the Sangro Valley Project. In M. Gillings, D. Mattingly and J. van Dalen (eds.), *Geographical Information Systems and Landscape Archaeology.* Mediterranean Landscape Archaeology 3. Oxford: Oxbow.

Martin, D.

1996   *Geographical Information Systems: Socio-economic Applications.* 2nd edn. London: Routledge.

Martlew, R.

1996   The contribution of GIS to the study of landscape evolution in the Yorkshire Dales, UK. In H. Kamermans and K. Fennema (eds.), *Interfacing the Past: Computer Applications and Quantitative Methods in Archaeology CAA95*, 293–96. Analecta Praehistorica Leidensia 28. Leiden: University of Leiden Press.

McDonald, W., and G.R. Rapp

1972   *The Minnesota Messenia Expedition: Reconstructing a Bronze Age Regional Environment.* Minneapolis: University of Minnesota Press.

McGlade, J.

1996   Whither GIS? Spatial technologies and the pursuit of archaeological knowledge. Paper presented at the 18th Annual Theoretical Archaeology Group Conference, Liverpool.

1997   GIS and Integrated Archaeological Knowledge Systems. In M. North and I. Johnson (eds.), *Archaeological Applications of GIS: Proceedings of Colloquium II, UISPP XIIIth Congress.* Archaeological Methods Series 5. Sydney: Sydney University.

Mee, C., and H. Forbes

1997   *A Rough and Rocky Place: The Landscape and Settlement History of the Methana Peninsula, Greece.* Liverpool: Liverpool University Press.

Pickles, J.

1995   *Ground Truth: The Social Implications of Geographic Information Systems.* New York: Guilford Press.

1997   Tool or Science? GIS, Technoscience, and the Theoretical Turn. *Annals of the Association of American Geographers* 87(2): 363–72.

Rimmington, J.N.

in press   Soil geochemistry and artefact scatters in Boeotia, Greece. In M. Pasquinucci and F. Trément (eds.), *Non-destructive Techniques Applied to Landscape Archaeology.* Mediterranean Landscape Archaeology 4. Oxford: Oxbow.

Robinson, J.M., and E.B.W. Zubrow

in press   Between Spaces: Interpolation in Archaeology. In M. Gillings, D. Mattingly and J. van Dalen (eds.), *Geographical Information Systems and Landscape Archaeology.* Mediterranean Landscape Archaeology 3. Oxford: Oxbow.

Savage, S.H.

1990   Modelling the late archaic social landscape. In K.M.S. Allen, S.W. Green and E.B.W. Zubrow (eds.), *Interpreting Space: GIS and Archaeology*, 330–55. London: Taylor & Francis.

Schofield, J.

1993   Looking back with regret; looking forward with optimism: making more of surface lithic scatters. In N. Ashton and A. David (eds.), *Stories in Stone: Lithic Studies Society Occasional Paper 4*, 90–98. London: Lithic Studies Society.

1999   Reflections on the future of surface lithic artefact study in England *(this volume)*.

Schofield, J. (ed.)

1990   *Interpreting Artefact Scatters.* Oxford: Oxbow.

Simmons, I.G.

1993   *Interpreting Nature: Cultural Constructions of the Environment.* London: Routledge.

Van Leusen, P.M.

1995   GIS and archaeological resource management: a European agenda. In G. Lock and Z. Stancic (eds.), *Archaeology and Geographical Information Systems: A European Perspective*, 27–42. London: Taylor & Francis.

Verhagen, P.

1996   The use of GIS as a tool for modelling ecological change and human occupation in the Middle Arhus Valley (S.E. Spain). In H. Kamermans and K. Fennema (eds.), *Interfacing the Past: Computer Applications and Quantitative Methods in Archaeology CAA95*, 317–24. Analecta Praehistorica Leidensia 28. Leiden: University of Leiden Press.

Verhagen, P., J. McGlade, R. Risch and S. Gili

1995   Some criteria for modelling socio-economic activities in the Bronze Age of south-east Spain. In G. Lock and Z. Stančič (eds.), *The Impact of Geographic Information Systems on Archaeology: A European Perspective*, 187–210. New York: Taylor & Francis.

Wansleeben, M., and L. Verhart

1997   Geographical Information Systems: Methodological progress and theoretical decline? *Archaeological Dialogues* 4(1): 53–70.

Wheatley, D.

1993   Going over old ground: GIS, archaeological theory and the act of perception. In J. Andresen, T. Madsen and I. Scollar (eds.), *Computing the Past: Computer Applications and Quantitative Methods in Archaeology*, 133–38. Aarhus: Aarhus University Press.

1996   Between the lines: the role of GIS-based predictive modelling in the interpretation of extensive survey data. In H. Kamermans and K. Fennema (eds.), *Interfacing the Past: Computer Applications and Quantitative Methods in Archaeology CAA95*, 275–92. Analecta Praehistorica Leidensia 28. Leiden: University of Leiden Press.

Whitelaw, T.M.
1991 Investigations at the Neolithic Sites of Kephala and Paoura. In J.F. Cherry, J.L. Davis and E. Mantzourani, *Landscape Archaeology as Long-Term History: Northern Keos in the Cycladic Islands from Earliest Settlement until Modern Times*, 199–26. Los Angeles: University of California.

Witcher, R.
in press GIS and Landscapes of Perception. In M. Gillings, D. Mattingly and J. van Dalen (eds.), *Geographical Information Systems and Landscape Archaeology*. Mediterranean Landscape Archaeology 3. Oxford: Oxbow.

Wright, D.J., M.F. Goodchild and J.D. Proctor
1997 Demystifying the Persistent Ambiguity of GIS as 'Tool' versus 'Science'. *Annals of the Association of American Geographers* 87(2): 346–62.

Zubrow, E.B.W.
1994 Knowledge representation and archaeology: a cognitive example using GIS. In C. Renfrew and E.B.W. Zubrow (eds.), *The Ancient Mind: Elements of Cognitive Archaeology*, 107–18. Cambridge: Cambridge University Press.